Mathematics and Computer Science II

CWI Monographs

Managing Editors

J.W. de Bakker (CWI, Amsterdam)
M. Hazewinkel (CWI, Amsterdam)
J.K. Lenstra (CWI, Amsterdam)

Editorial Board

W. Albers (Maastricht)
P.C. Baayen (Amsterdam)
R.T. Boute (Nijmegen)
E.M. de Jager (Amsterdam)
M.A. Kaashoek (Amsterdam)
M.S. Keane (Delft)
J.P.C. Kleijnen (Tilburg)
H. Kwakernaak (Enschede)
J. van Leeuwen (Utrecht)
P.W.H. Lemmens (Utrecht)
M. van der Put (Groningen)
M. Rem (Eindhoven)
A.H.G. Rinnooy Kan (Rotterdam)
M.N. Spijker (Leiden)

Centrum voor Wiskunde en Informatica
Centre for Mathematics and Computer Science
P.O. Box 4079, 1009 AB Amsterdam, The Netherlands

The CWI is a research institute of the Stichting Mathematisch Centrum, which was founded on February 11, 1946, as a nonprofit institution aiming at the promotion of mathematics, computer science, and their applications. It is sponsored by the Dutch Government through the Netherlands Organization for the Advancement of Pure Research (Z.W.O.).

CWI Monograph 4

Mathematics and Computer Science II

Fundamental contributions in the Netherlands since 1945

edited by
M. Hazewinkel
J. K. Lenstra
L. G. L. T. Meertens

1986

North-Holland
Amsterdam · New York · Oxford · Tokyo

© Centre for Mathematics and Computer Science, 1986
All rights reserved. No part of this publication may be reproduced, stored in a retrieval system, or transmitted, in any form or by any means, electronic, mechanical, photocopying, recording or otherwise, without the prior permission of the copyright owner.

ISBN: 0 444 70122 2

Publishers:
Elsevier Science Publishers B.V.
P.O. Box 1991
1000 BZ Amsterdam
The Netherlands

Sole distributors for the U.S.A. and Canada:
Elsevier Science Publishing Company, Inc.
52 Vanderbilt Avenue
New York, N.Y. 10017
U.S.A.

Cover: Tobias Baanders

Printed in the Netherlands

Preface

The year 1986 is marked by the 40th anniversary of the Mathematical Centre, and also of its research institute, the Centre for Mathematics and Computer Science (CWI), formerly also known as 'Mathematical Centre'. The founders of the Mathematical Centre, J.G. van der Corput, D. van Dantzig, J.F. Koksma, H.A. Kramers, M.G.J. Minnaert and J.A. Schouten, felt that the mathematical sciences should contribute to the rebuilding of the Netherlands after World War II. Scientific development in the Low Countries had come to a halt during the war, and a concentrated effort would be required for making up the arrears. The six founders held the firm conviction that their aims would be furthered best by using the problems arising in the application of mathematical methods and results to practical problems as a source of inspiration, while maintaining a strong focus on fundamental research. It was from this conviction that the plans were forged, already during the war, that led to the founding of the Mathematical Centre on 11 February 1946.

Forty years later, their ideas have proven in no way outdated. Now that the scars afflicted by World War II are largely healed, practical problems still inspire promising avenues of research, and the most significant contributions to the practical applicability of the mathematical sciences originate from fundamental research. While, only a few decades ago, we would have been hard put to give more than a few convincing recent examples of the latter, nowadays they abound, and the reader will have no difficulty to supply many themselves. It appears that the 'depth investment' has paid off, and continues to do so. The above does not purport to suggest that the actual applicability of fundamental results that are applicable in principle is a small matter, which is, as it were, solved as soon as the results have been obtained. On the contrary: translating these results into a tool that can be routinely applied by the 'workers in the field' often still takes a substantial and conscious development effort.

What, then, are the hallmarks of fundamental research that distinguish it from its application-bound counterpart, and how can its surprising efficacy be explained? Surely, these two questions are related, and to answer one is to come at least halfway to answering the other. These questions have occupied many philosophers of science, and any answers we may offer are at best partial: as with good art, it is easier to recognize a fundamental approach than to characterize it. The most important asset, perhaps, of a successful fundamental researcher, is the ability to identify the essential aspects of a problem and to concentrate on those, leaving out most of the problem statement. If necessary, this process is repeated. More likely than not, the original problem is thereby transformed beyond recognizability. The abstract problem statement thus obtained will often be found to be tied to the web of established mathematical theory; if not in the sense that it has already been studied, then at least in its similarity to other problems — usually stemming from entirely unrelated sources — which may suggest potential methods of attack. While there is, of course, no guarantee that a solution of the transformed problem is transferable to the original one, by the very meaning of the word 'essential' we may at least expect that insights obtained in the abstract problem setting will shed light on the concrete version, which in its full-fledged form may appear quite impervious to the strongest forms of direct attack.

Not only do we feel that the success of fundamental research — in spite of our attempts at an explanation — remains surprising, the last decades have also shown an astonishingly fruitful applicability of various mathematical fields to formerly quite different and already well-developed other mathematical fields: in particular, but not alone, of algebra and of geometry (and of these two to each other). But completely new theories have also sprung up, sometimes against all preconceived likelihood, offering a handle on problems that were thus far deemed intractable. Only through the fundamental approach, through a concentration on the bare essence of a problem, have these important advances been possible.

In view of the pre-eminent role played by fundamental research in the efforts of the Mathematical Centre, it was only fitting that a symposium on fundamental contributions to Mathematics and Computer Science in the Netherlands, over the period of its existence, was organized to commemorate its fortieth anniversary. Papers have been sollicited that, although they cannot, obviously, touch on all of the areas of research that have been covered, together are indicative of the breadth and depth of the subjects studied. Although not made explicit in the title, at least a relationship to research conducted at the CWI was a must. This may seem an undue restriction, but it is fair to say that there is little thereby excluded — one of several notable exceptions is the field of mathematical logic, in which there is nevertheless a strong Dutch contribution, in particular in relationship to foundational issues. In each of the contributions to this monograph, which constitutes the proceedings of the symposium mentioned above, both the practical source of the problems studied and the relevance of the results to such problems are apparent. This was not specifically sought, but neither, do we think, is it a coincidence. We feel that these papers can prove a source of inspiration to researchers in Mathematics and Computer Science, whatever their field of specialization. In particular, we think that read together they offer excellent examples of the merits of an open-minded and broad approach to hard problems. Such approaches often leave the obvious roads, which would have led to an impasse, and turn into unexpected directions.

The symposium was held in Amsterdam on 6 and 7 October 1986. It is our pleasure to thank all those who helped to realize this monograph in what was, realistically speaking, too short a period of time: the authors; the typing staff, and in particular Ms. C.J. Swagerman; the computer typesetting group; and Mr. W.A.M. Aspers, the desk editor. We also wish to extend our thanks to all those who contributed in organizing or otherwise to the symposium.

M. Hazewinkel
J.K. Lenstra
L.G.L.T. Meertens

Contents

The numerical solution of partial differential equations   1
   A.O.H. Axelsson
Dynamics in bio-mathematical perspective   23
   O. Diekmann
The arch-enemy attacked mathematically   51
   L. de Haan
Process algebra: specification and verification in bisimulation semantics   61
   J.A. Bergstra, J.W. Klop
Codes from algebraic number fields   95
   H.W. Lenstra, Jr.
Infinite-dimensional normed linear spaces and domain invariance   105
   J. van Mill
Geometric methods in discrete optimization   111
   A. Schrijver
Archirithmics or algotecture?   139
   P.M.B. Vitányi

The Numerical Solution of Partial Differential Equations

A.O.H. Axelsson
Department of Mathematics, University of Nijmegen
Toernooiveld, 6525 ED Nijmegen, The Netherlands

We shall describe a few aspects of the numerical solution of partial differential equations which are related to work done by the numerical group at the Centre for Mathematics and Computer Science (CWI) in Amsterdam.

1. INTRODUCTION

Many problems in, for example, physics, biophysics and technology are described by partial differential equations (PDEs). Often, these equations are of second order; we distinguish elliptic, parabolic and hyperbolic equations. Examples of such equations are respectively given by

$$-\sum_{i=1}^{d}[\frac{\partial}{\partial x_i}(a_i\frac{\partial u}{\partial x_i}) + b_i\frac{\partial u}{\partial x_i}] = f, \; a_i > 0, \; \vec{x} \in \Omega \in \mathbb{R}^d \tag{1.1}$$

$$\frac{\partial u}{\partial t} = \sum_{i=1}^{d}\frac{\partial}{\partial x_i}(a_i\frac{\partial u}{\partial x_i}) + f, \; a_i > 0, \; \vec{x} \in \Omega, \; t \geq 0 \tag{1.2}$$

$$\frac{\partial^2 u}{\partial t^2} = \sum_{i=1}^{d}\frac{\partial}{\partial x_i}(a_i\frac{\partial u}{\partial x_i}) + f, \; a_i > 0, \; \vec{x} \in \Omega, \; t > 0 \tag{1.3a}$$

$$\frac{\partial u}{\partial t} + \sum_{i=1}^{d}b_i\frac{\partial u}{\partial x_i} = f, \; \vec{x} \in \Omega, \; t > 0, \tag{1.3b}$$

The equations (1.3a) and (1.3b) are both of hyperbolic type. Here, Ω is a bounded (but not always simply connected) or sometimes an unbounded domain in $\mathbb{R}^d$, $d = 1$, 2 or 3. Along the boundary $\partial\Omega$ of Ω boundary conditions are prescribed, e.g., $u = g_0$ (Dirichlet) or $\partial u/\partial n = g_1$ (Neumann) or mixed boundary conditions: $u = g_0$ along Γ_1, $\partial u/\partial n = g_1$ along Γ_2 where $\Gamma_1 \cup \Gamma_2 = \partial\Omega$ ($\Gamma_1 \cap \Gamma_2 = \emptyset$).

Equation (1.1) represents a diffusion equation, or convection-diffusion equation, whereas (1.2) is a nonstationary diffusion equation, or heat equation; the equations (1.3a,b) represent wave equations. Solutions of (1.3a,b) possess a bounded domain of dependency, that is, the solution $u(\vec{x}, t)$ depends on data prescribed in a bounded domain, whereas solutions of (1.1) and (1.2) depend on data prescribed at all points. In the case (1.3a,b) with vanishing source function f, the energy (kinetic plus potential energy) is constant for all t, while in the case (1.2) (with $f\equiv 0$) the energy decreases exponentially with t.

Problems of mixed type also occur; for example, equation (1.1) with $a_i > 0$, $i = 1, \ldots, d$ on $\Omega_1 \subset \Omega$ and with one or two of the coefficients $a_i \leq 0$ on $\Omega_2 = \Omega \setminus \Omega_1$. Thus, the equation is of elliptic type on Ω_1 and of hyperbolic type on Ω_2.

Equations such as

$$\sum_{i=1}^{d} \left[-\epsilon \frac{\partial^2 u}{\partial x_i} + b_i \frac{\partial u}{\partial x_i} \right] = f$$

are elliptic, however, for small values of ϵ (singularly perturbed problems), the equation exhibits a hyperbolic character rather than elliptic.

In most cases, partial differential equations cannot be solved by analytical means, and therefore, numerical techniques are employed to obtain approximate solutions. Important discretization techniques are:
(i) finite difference methods;
(ii) finite elements and finite volume methods;
(iii) boundary element methods;
(iv) spectral methods.

In the numerical analysis of these methods there are three crucial problems:
(a) stability of the numerical solution, i.e., does the solution depend continuously on the given data;
(b) discretization errors, i.e., the difference of the continuous and discrete solution in some norm;
(c) the efficient solution of the algebraic systems to be solved when applying the discretization method.

A discussion of these problems can be found in the text books [2], [3], [13], [16], [38], [39], [43], and [47].

The finite difference discretization of equation (1.1) is of the form

$$-\sum_{i=1}^{d} D_i^+ \left(b_i(\vec{x} + \tfrac{1}{2} h \vec{e}_i) D_i^- u(\vec{x}) \right) = f(\vec{x}), \quad \vec{x} \in \Omega_h,$$

where $\Omega_h \subset \Omega$ denotes a grid with meshes of size h, and D_i^-, D_i^+ denote the backward and forward difference operators defined by

$$D_i^- u(\vec{x}) = \frac{1}{h} [u(\vec{x}) - u(\vec{x} - h\vec{e}_i)]$$

$$D_i^+ u(\vec{x}) = \frac{1}{h} [u(\vec{x} + h\vec{e}_i) - u(\vec{x})]$$

with $\vec{e}_i$ is the ith unit vector in $\mathbb{R}^d$.

Finite element and finite volume techniques, which use numerical quadrature, also generate difference equations, however, these equations are of more general nature than in the case of finite difference methods. Finite element and finite volume methods are more easily applied in the case of curved boundaries.

For elliptic equations, the stability does not offer problems: for difference methods a discrete maximum principle can be proved, and for finite elements methods there is a coercive bilinear form from which it can be proved that the stiffness matrix, associated to the particular finite element method, is positive definite. Discretization errors are of order $O(h^2)$ as $h \to 0$ for difference methods and of arbitrarily high order if an appropriate finite element method is chosen (provided that the solution is sufficiently smooth). In passing we remark that for singular problems and for problems with corners there are also suitable methods available. The number of equations to be solved when applying (i) or (ii) can be very large (in 3-dimensional problems one easily has systems with $10^5 - 10^6$ unknowns); however, the matrix of coefficients is, fortunately, very sparse (only a relatively small number of elements does not vanish). This feature should be exploited in solving the algebraic systems. In the case of the methods (iii) and (iv) the number of equations is smaller, but the matrix of coefficients is less sparse. In this contribution we shall concentrate on methods of type (i) and (ii).

In solving the algebraic system $Ax = b$ one may use direct and iterative methods. Direct methods furnish the solution, (with rounding errors), in a finite number of steps, whereas iterative methods yield a sufficiently accurate approximation after a number of steps which usually depends on the condition number of the matrix A. Some methods, such as the conjugate gradient method, are hybrid methods, that is they are both of direct and of iterative type; however, they are mostly applied in an iterative fashion, because they furnish a sufficiently accurate solution in a much smaller number of steps than required for reaching the 'exact' solution.

Direct methods usually employ a factorization $L \cdot U$ of the matrix A; here, L denotes a lower triangular and U an upper triangular matrix. Under certain conditions, related to the order of the grid points, the matrices L and U are also sparse, but, generally, to a less extent than the matrix A itself. For example, in 3-dimensional problems, defined on a cube with n^3 grid points say, it can be shown that the computational complexity is $O(n^6)$ and the necessary storage $O(n^4)$. These figures correspond to a special nested dissection order (cf. [31] and [3]). When applying the more conventional (and 'natural') ordering of the grid points, these figures are, respectively, $O(n^7)$ and $O(n^5)$.

The matrix A itself requires, at most, storage of $O(n^3)$, typically $7n^3$, but sometimes, in the case of constant coefficients and uniform grids, much less.

Certain iterative methods, so-called preconditioned conjugate gradients methods (PCG methods), have, under certain conditions, a computational complexity of $O(n^{3.5})$ and require storage of $O(n^3)$ as $n \to \infty$. For so-called multigrid methods, these figures are, under certain conditions, $O(n^3 \log n)$ and $O(n^3)$, respectively.

In the case of large scale problems, $n = 50, \ldots, 100$ say, it is even on super computers necessary to limit storage and computational effort drastically. Direct methods are not feasible for 3-dimensional problems. For example, if $n = 64$ and using the natural order of the grid points, then about $10^9 = 1000$ M has to be stored. (On the CYBER 205 of SARA there is 1 M storage available, so that one should resort to background storage. However, the number of necessary input/output (I/O) transfers is *at least* 1000 M which is sometimes much more expensive than the computation time. Moreover, the computation time itself in this problem (on the CYBER 205) is at least $2 \cdot n^7 / 50 \cdot 10^{-6}$ seconds, that is about 45 CPU hours!)

In this lecture we shall give a survey of the multigrid method; next, we shall discuss a few aspects of the numerical solution of time-dependent PDEs, and finally, some other aspects, such as the use of supercomputers, which are important in the numerical solution of PDEs will be discussed.

2. THE MULTIGRID METHOD

We shall present the multigrid method, taking as a starting point the defect-correction method and the variational formulation of (1.1),

$$a(u,v) = (f,v) \quad \forall v \in H_0^1(\Omega),$$

where

$$a(u,v) = \sum_{i=1}^{d} \int_{\Omega} a_i \frac{\partial u}{\partial x_i} \frac{\partial v}{\partial x_i} d\Omega$$

and $H_0^1(\Omega)$ is the Sobolev space of order 1. We assume $\Omega \subset \mathbb{R}^2$ to be a simply connected polygonal region and the solution $\hat{u}$ of (1.1) to be sufficiently smooth. A sequence of triangular grids $\Omega_h^{(1)}, \Omega_h^{(2)}$ is constructed as shown in figure 2.1.

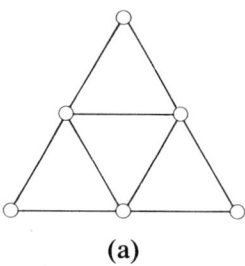
(a)

FIGURE 2.1
Finite elements for the grids
$\Omega_h^{(1)}$ and $\Omega_h^{(2)}$ respectively

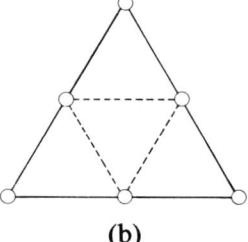
(b)

In figure (2.1a) we use piecewise linear basis-functions (λ_i) and in (2.1b) piecewise quadratic basis-functions (ϕ_i). Notice that for all h the nodal points coincide for the two grids. Hence the order of the corresponding finite-element matrices are equal. Thus, we obtain finite-element matrices $K_h^{(i)}$ and solutions $\hat{U}_h^{(i)}$, $i = 1, 2$ of

$$K_h^{(1)} \hat{U}_h^{(1)} = f_h^{(1)}, \quad f_{h,i}^{(1)} = (f, \lambda_i),$$

and

The numerical solution of partial differential equations

$$K_h^{(2)} \hat{U}_h^{(2)} = f_h^{(2)}, \quad f_{h,i}^{(2)} = (f, \phi_i)$$

respectively.

Notation: we use small letters for the corresponding approximating finite element functions:

$$\hat{u}_h^{(1)} = \sum \hat{U}_{h,i}^{(1)} \lambda_i;$$

and

$$\hat{u}_h^{(2)} = \sum \hat{U}_{h,i}^{(2)} \phi_i,$$

where the summation is meant over all nodal points. For the approximating functions we know the discretization errors

$$\|\hat{u} - \hat{u}_h^{(1)}\| = O(h^{\nu_1}), \quad \|\hat{u} - \hat{u}_h^{(2)}\| = O(h^{\nu_2}) \text{ for } h \to 0,$$

where $\nu_2 \geqslant \nu_1$. (Under conditions we have $\nu_1 = 1$, $\nu_2 = 2$). For the solution of the higher-order method (with, in general, a somewhat more complicated matrix) we shall use the lower-order method, for which the matrix has a more simple structure. This difference in structure makes it cheaper to solve the system with $K_h^{(1)}$ than the system with $K_h^{(2)}$.

2.1. The defect-correction method

Using a defect-correction method, one first solves an initial approximation

$$K_h^{(1)} U_h^{(0)} = f_h^{(1)} \quad \text{(i.e. } U_h^{(0)} = \hat{U}_h^{(1)}\text{)},$$

and then a correction from the equation

$$K_h^{(1)} \delta_h^{(1)} = -r_h^{(2)},$$

where the residual $r_h^{(2)}$ is defined by

$$r_h^{(2)} = K_h^{(2)} U_h^{(0)} - f_h^{(2)}.$$

Finally we set

$$u_h^{(1)} = \tilde{u}_h^{(0)} + \delta_h^{(1)}.$$

where $\tilde{u}_h^{(0)}$ is a function, written as a linear combination of the functions ϕ_i, i.e.

$$\tilde{u}_h^{(0)} = \sum U_{h,i}^{(0)} \phi_i.$$

We want to estimate the error $\|\hat{u} - u_h^{(1)}\|_1$. We have

$$K_h^{(1)}(U_h^{(1)} - \hat{U}_h^{(2)}) = K_h^{(1)}(U_h^{(1)} - U_h^{(0)}) + K_h^{(1)}(U_h^{(0)} - \hat{U}_h^{(2)}) \quad (2.1)$$
$$= (K_h^{(1)} - K_h^{(2)})(U_h^{(0)} - \hat{U}_h^{(2)}).$$

From the fact that the solution is smooth, the existence of sufficiently smooth functions $e^{(i)} \in H^1(\Omega)$ follows, $e^{(i)}$ independent of h such that

$$\hat{u}_h^{(1)} - \hat{u} = h^2 e^{(1)} + O(h^3), \quad h \to 0,$$
$$\hat{u}_h^{(2)} - \hat{u} = h^3 e^{(2)} + o(h^3), \quad h \to 0$$

in the sense that $\|\hat{u}_h^{(1)} - \hat{u} - h^2 e^{(1)}\|_0 = O(h^3)$ and $\|\hat{u}_h^{(1)} - \hat{u} - h^2 e^{(1)}\|_1 = O(h^2)$, etc. (cf. e.g. [42]).

Thus, we have

$$u_h^{(0)} - \hat{u}_h^{(2)} = \hat{u}_h^{(1)} - \hat{u} - (\hat{u}_h^{(2)} - \hat{u}) = h^2 e^{(1)} + O(h^3).$$

From the fact that $e^{(1)}$ is sufficiently smooth it also follows that

$$(K_h^{(1)})^{-1}(K_h^{(1)} - K_h^{(2)})e^{(1)} = O(h). \tag{2.2}$$

Hence, from (2.1) and (2.2) follows

$$\|u_h^{(1)} - \hat{u}_h^{(2)}\|_1 = O(h^2)$$

and finally

$$\|u_h^{(1)} - \hat{u}\|_1 \leq \|u_h^{(1)} - \hat{u}_h^{(2)}\|_1 + \|\hat{u}_h^{(2)} - \hat{u}\|_1$$
$$= O(h^{\min(2, v_2)}) = O(h^2).$$

Thus, by solving *two* times a system with $K_h^{(1)}$, we can obtain the order of accuracy of the quadratic problem. This, or similar, problems have been analyzed by HEMKER in [23], and, in connection with the convection-diffusion equation in [24]. If one wants to use a PCG-method for the solution of $K_h^{(2)} U_h^{(2)} = f_h^{(2)}$, there is an alternative method in deriving a preconditioning matrix from $K_h^{(1)}$, but using it for the solution of the system with $K_h^{(2)}$ (see [3]).

2.2. The multigrid method

Let us first consider two grids $\Omega_{h_2} \subset \Omega_{h_1}$, $h_1 < h_2$, in a sequence of grids $\{\Omega_{h_i}\}$. A multigrid method consists of two stages: 'smoothing' and 'correction'.

Stage 1: *smoothing*. On the fine grid Ω_{h_1} a number of relaxation sweeps is made in order to reduce the more rapidly oscillating components in the present residual. This can be done e.g. by a Gauss-Seidel-, a Gauss-Jacobi- or semi-iterative Chebychev-method (we return later to this method). This 'smoothing' results in a new approximation $u_{h_1}^{(0)}$ and a residual $r_{h_1}^{(0)} = K_{h_1} u_{h_1}^{(0)} - f_{h_1}$ on Ω_{h_1}.

Stage 2: *correction*. Now we compute a correction $\delta_{h_1}^{(1)}$, that satisfies

$$K_{h_1} \delta_{h_1}^{(1)} = -r_{h_1}^{(0)} \tag{2.3}$$

and a new approximation $u_{h_1}^{(1)} = u_{h_1}^{(0)} + \delta_{h_1}^{(1)}$. The costs to find the exact solution to (2.3) are equal to those for the solution of the original system. This is too large of course. Therefore, the fundamental idea in the multigrid method is to compute the correction on the coarser grid, where the costs are much less. Notice that the approximation of $\delta_{h_1}^{(1)}$ obtained on the coarser grid can be a good approximation because the right-hand-side, $r_{h_1}^{(0)}$, consists mainly of smooth components and, therefore, can be approximated well on the coarser grid.

In order to formulate the correction stage as a defect-correction process, we define restriction and prolongation operators between the different grids. The restriction operator is defined as follows. For any u defined on Ω_{h_1}, let $I_{h_1}^{h_2} u$ be a function on Ω_{h_2}, which satisfies

$$I_{h_1}^{h_2} u(N_j) = u(N_j), \quad N_j \in \Omega_{h_2},$$

and is defined further by interpolation on Ω_{h_2}, for all interior points. As an alternative, the nodal values of $I_{h_1}^{h_2} u$ may also be obtained as a weighted mean of nodal values of u at neighbouring points.

For the prolongation operator we simply use the interpolation operator: for any u defined on Ω_{h_2}, we define $I_{h_2}^{h_1} u$ as a function on Ω_{h_1} determined by

$$I_{h_2}^{h_1} u(N_j) = u(N_j), \quad N_j \in \Omega_{h_1}$$

and by interpolation for the interior points. Notice that for finite elements there exists a natural definition based on the global character of the approximating functions; each function can be found in the same Sobolov space $H^1(\Omega)$.

The correction stage of the multigrid method is now as follows: Solve on Ω_{h_2} the equation

$$K_{h_2} \delta_{h_2}^{(0)} = -I_{h_1}^{h_2} r_{h_1}^{(0)}$$

and let

$$u_{h_1}^{(1)} = u_{h_1}^{(0)} + I_{h_2}^{h_1} \delta_{h_2}^{(0)}$$

be the correction on Ω_{h_1}. Then we have

$$u_{h_1}^{(1)} = u_{h_1}^{(0)} - I_{h_2}^{h_1} K_{h_2}^{-1} I_{h_1}^{h_2} r_{h_1}^{(0)}$$

i.e.

$$r_{h_1}^{(1)} = (I_{h_1}^{h_1} - K_{h_1} I_{h_2}^{h_1} K_{h_2}^{-1} I_{h_1}^{h_2}) r_{h_1}^{(0)}$$

(where $I_{h_1}^{h_1}$ is the identity operator on Ω_{h_1}). The rate of convergence depends on the magnitude of the right-hand-side. Because $r_{h_1}^{(0)}$ is a 'smoothed' residual, this decrease mainly depends on the approximation properties, i.e. we can expect

$$(I_{h_1}^{h_1} - K_{h_1} I_{h_2}^{h_1} K_{h_2}^{-1} I_{h_1}^{h_2}) r_{h_1}^{(0)} = K_{h_1} (K_{h_1}^{-1} I_{h_1}^{h_1} - I_{h_2}^{h_1} K_{h_2}^{-1} I_{h_1}^{h_2}) r_{h_1}^{(0)}$$

to be $O(h_1)$ for $h_1, h_2 \to 0$.

(Compare this with the expression (2.2) for the defect-correction method.) If $r_{h_1}^{(1)}$ is not sufficiently small, we should repeat the whole process, that means first 'smoothing' and then 'correcting', but now starting with $r_{h_1}^{(1)}$ and $u_{h_1}^{(1)}$ instead of $r_{h_1}^{(0)}$ and $u_{h_1}^{(0)}$. This all can be iterated/repeated again until the residual is sufficiently small. Thus far we only considered the fundamental two-level step of the multigrid method. The full multigrid method uses a whole sequence of grids in the following way. Solving the equation (2.3) on Ω_{h_2}, we use again the fundamental two-level step, but now on the grids Ω_{h_2} and Ω_{h_3}, with $h_3 > h_2$. In this way we can continue until we reach a sufficiently coarse grid, where a direct method is the most efficient way for solving the equation.

For an illustration of the effect of the smoothing stage we consider the following model problem:

$$\Delta u = f \text{ on } \Omega$$

$$u = 0 \text{ on } \delta\Omega$$

with $\Omega = [0,1]^d$; Ω_h is a regular grid and Δ is approximated by the difference operator Δ_h. The eigenvalues and eigenfunctions of $\Delta_h v = \mu v$ are

$$S_h = \{\mu\} = \left\{ \sum_{i=1}^{d} \left(\frac{2 \cdot \sin \frac{\pi p_i h}{2}}{h} \right)^2 \right\}, \quad p_i \in \{1, 2, \ldots, n\},$$

the spectrum, and

$$v_p = \prod_{i=1}^{d} \sin(p_i \pi k_i h), \quad k_i \in \{1, 2, \ldots, n\}$$

the eigenfunctions, where $n = \frac{1}{h} - 1$ and $n = 2^m$.

Notice that with $h_1 = h$ and $h_2 = 2h$, Δ_{h_2} on the grid Ω_{h_2} has the same eigensolutions, but with $k_i, p_i \in \{1, 2, \ldots, n/2\}$, where $n/2 = 2^{m-1}$. The eigenvalues of Δ_{h_1} that are not present on Ω_{h_2} (those with rapidly oscillating eigenfunctions) are found in the interval $(dh^{-2}, 4dh^{-2})$. This interval has a 'condition-number' $\kappa = 4$, independent of h.

With a Gauss-Jacobi relaxation

$$U_h^{(l)} = U_h^{(l-1)} - \tau(K_h U_h^{(l-1)} - f_h), \quad l = 1, \ldots, \nu$$

we obtain the residuals

$$r_h^{(l)} = (I - \tau K_h) r_h^{(l-1)} = (I - \tau K_h)^l r_h^{(0)}.$$

If $r_h^{(0)} = \Sigma \alpha_p v_p$, we obtain $r_h^{(l)} = \Sigma(1 - \tau \mu_p)^l \alpha_p v_p$.

We choose τ such that the components corresponding to eigenvalues in the interval $(dh^{-2}, 4dh^{-2})$ will be damped most.

We see that we better can choose variable τ's, i.e.

$$U_h^{(l)} = U_h^{(l-1)} - \tau_l(K_h U_h^{(l-1)} - f_h), \quad l = 1, \ldots, \nu$$

and
$$r_h^{(1)} = \sum q_l(\mu_p)\alpha_p v_p$$
where $q_l(\mu) = \Pi_{l=1}^{\nu}(1-\tau_l\mu)$. The best choice of τ_l is such that we obtain a polynomial q_l^* for which
$$\max|q_l^*(\mu)| \leq \min_{q_l} \max|q_l(\mu)|,$$
where the maximum is taken over the interval $(a,b)=(dh^{-2}, 4dh^{-2})$. It is well-known that
$$q_l^*(\mu) = T_l\left(\frac{2\mu-(b+a)}{b-a}\right)/T_l\left(\frac{b+a}{b-a}\right)$$
and
$$\max|q_l^*(\mu)| = 1/T_l\left(\frac{b+a}{b-a}\right).$$

Here, T_l is the Chebychev polynomial, normalized on the interval $[0,1]$, with $T_l(0)=1$.

Now, the asymptotic convergence factor is
$$\rho = \frac{1-1/\sqrt{\kappa}}{1+1/\sqrt{\kappa}} = \frac{1}{3}.$$

This means a small (mean) convergence factor per relaxation sweep. Such a smoothing method has been proposed by VAN DER HOUWEN and SOMMEIJER in [32].

In general a fixed number of 'smoothing' relaxations is performed on each level of discretization (except for the coarsest Ω_{h_m}).

The resulting approximation on Ω_{h_m} is used for a correction on $\Omega_{h_{m-1}}$. Now a choice can be made. Either we can do a new smoothing step on the grid $\Omega_{h_{m-1}}$ and make a new correction on the coarser grid before we return to the finer level, or we can return to the finer level immediately. In both cases we can do a few smoothing steps before we return to the finer level (this is called post-smoothing). On the level h_{m-1} we have the same choice etc. If the first choice is made on all levels we speak of a W-cycle; the other choice is called a V-cycle.

Notice that the computational costs of a fixed number of relaxation sweeps on level h is $O(N_h)$, where N_h is the number of unknowns on the h-level. Let $h_i = 2^i h_0$, where $h = h_0$ denotes the finest level. With the assumption of a fixed number ν relaxation sweeps on each level for a single cycle, and assuming that the costs of a relaxation sweep (i.e. matrix-vector multiplication plus a vector sum) is Kh^{-d} on $\Omega_h \subset \mathbb{R}^d$, we can compute the computational complexity of a cycle as follows.

We denote by μ the number of times that we return to a coarser grid for the computation of a correction stage. Thus, $\mu=1$ for a V-cycle and $\mu=2$ for a

W-cycle. Let $Q(h)$ be the costs on the grid Ω_h, then, for $\mu<2^d$, we find

$$Q(h) = \nu K h^{-d} + \mu Q(2h)$$
$$= \nu K h^{-d} + \mu \nu K(2h)^{-d} + \mu^2 Q(4h)$$
$$\leq \nu K h^{-d}[1 + \mu/2^d + (\mu/2^d)^2 + \cdots]$$
$$= \frac{2^d}{2^d - \mu} \cdot \nu K h^{-d}.$$

Hence, the costs per cycle are $O(h^{-d})$ if $\mu<2^d$.

Under certain conditions one can prove that a fixed number of cycles, or in any case at most $O(\log h^{-1})$ cycles are sufficient to attain an iteration error that is less than the discretization error $O(h^2)$. This means that the complexity of the method is $O(N_h)$ or $O(N_h \log N_h)$, i.e. the complexity of the method is proportional (or almost proportional) to the number of the unknowns. Hence the complexity of the method is (almost) optimal.

The idea of a multigrid method was originally described in some papers by FEDORENKO [20], BACHVALOV [11] and ASTRACHANCEV [1]. Later BRANDT [15] has improved the method and emphasized its practical value. For some proofs on the convergence and the complexity of the method, see BANK and DUPONT [12], BRAESS [14], HACKBUSH [22], HEMKER [25], MCCORMICK and RUGE [17], NICOLAIDES [40] and WESSELING [51]. The smoothness of the solution has inference on the rate of convergence of the multigrid method. Some results concerning more robust multigrid methods, where the smoothness of the solution has less influence, can be found in DENDY [19] and HEMKER et al. [26].

One kind of robust multigrid methods uses particular relaxations as a smoother. These relaxations use a preconditioning matrix based on approximate factorization of the given matrix K_h. The most robust methods can be those which make use of a certain block-matrix structure of K_h, e.g. a partitioning in tridiagonal blocks. For recent reports see [37] and [4].

Other versions of multigrid methods exist that are based on projection on subspaces, see [3] and references herein and also YSERENTANT [53]. For such methods the smoothness of the solution has little influence.

More recently, multigrid methods have also been applied to non-symmetric systems, see e.g. DE ZEEUW and VAN ASSELT [54]. For recent successful results obtained with the application of multigrid methods to the Euler equations for compressible inviscid flows (computational fluid dynamics) see HEMKER and SPEKREIJSE [27,28].

3. TIME-DEPENDENT PARTIAL DIFFERENTIAL EQUATIONS

We shall first describe the so-called method of lines in the numerical solution of time-dependent PDEs and the stability and order of convergence connected with this method. Then we shall examine a so-called global method whereby the time variable is treated in the same way as a space variable. Finally we shall discuss the practically important problem of how to solve in an efficient and feasible way the large systems of linear or nonlinear algebraic equations arising in each time step of an implicit time stepping method.

3.1. The method of lines

One of the most frequently applied methods in the numerical solution of non-stationary PDE problems is the method of lines. To illustrate the ideas behind this method we shall consider the parabolic equation

$$\frac{\partial u}{\partial t} = \Delta u + \vec{v} \cdot \nabla u + f(\vec{x}, t), \quad \vec{x} \in \Omega \subset \mathbb{R}^d, \ t > 0, \tag{3.1}$$

supplied with boundary conditions, say $u = g$ on $\partial \Omega$, and with the initial condition $u(\vec{x}, 0) = u_0(\vec{x})$, $\vec{x} \in \Omega$. The vector $v = (v_1, \ldots, v_d)^T$ is here supposed to be constant.

In the method of lines one usually discretizes all independent variables with the exception of one. In the case of equation (3.1) one thus usually discretizes the space variable, which comes to the replacement of the Laplace operator Δ and the gradient operator ∇ by appropriate finite difference or finite element operators on a grid Ω_h. This replacement yields a system

$$\frac{dU_h(t)}{dt} = A_h U_h(t) + F_h(t), \quad t > 0, \tag{3.2}$$

of coupled ordinary differential equations with t as independent variable. Here, $U_h(t)$ is a grid function composed of the approximations $U_{h,i}(t)$ to $U(\vec{x}_i, t)$ in the grid points $\vec{x}_i \in \Omega_h$. Further, $U_{h,i}(0) = U_0(\vec{x}_i)$ and the inhomogeneous term $F_h(t)$ consists of the values $f(\vec{x}_i, t)$ and contributions from the boundary function $g(t)$.

The eigenvalues of the matrix A_h, the approximation to the linear spatial operator $\Delta + v \cdot \nabla$, are located in the left complex halfplane, at least if the grid distance h is sufficiently small (typically $h \leq \frac{1}{2}|v|$). This means that the continuous time, semi-discrete problem (3.2), like the original problem (3.1), is stable with respect to perturbations in the given data.

The true solution U_h of (3.2) is called the semi-discrete solution of the original problem (3.1). Using stability properties of (3.2) one can prove convergence of $U_{h,i}(t)$ to $U(\vec{x}_i, t)$ as $h \to 0$. For the standard central difference approximation on a uniform grid one thus finds the $O(h^2)$ behaviour for the semi-discrete error $U(\vec{x}_i, t) - U_{h,i}(t)$ (see e.g. VERWER [49], THOMÉE [48]). Moreover, this result is valid uniformly in t (see e.g. AXELSSON [6,7] and the references there in).

Within the method of lines one distinguishes two stages in the numerical solution process. The first stage is the semi-discretization as outlined above. The second stage consists of the numerical integration in time of the resulting system of ordinary differential equations, in the present case, system (3.2). In the numerical solution of this system, it is a prerequisite to maintain the stability in order to prevent that small errors such as rounding errors and truncation errors will be accumulated without bound when time evolves.

The most frequently applied numerical methods for systems like (3.2) are step-by-step methods. A well-known representative from this class reads for system (3.2)

$$(I - (1 - \theta)kA_h)U_{h,k}(t + k) = (I + \theta k A_h)U_{h,k}(t) + \qquad (3.3)$$
$$k((1 - \theta)F_h(t + k) + \theta F_h(t)),$$

where $t = 0, k, 2k, \ldots$ and $U_{h,k}(0) = U_h(0)$; k is the time stepsize and θ is a free parameter satisfying $0 \leq \theta \leq 1$. If we suppose that the matrix A_h has a complete eigensystem it is a simple matter to prove that this method is stable under the stepsize restriction

$$k \leq \frac{2}{2\theta - 1} \min_i \{ \frac{\text{Re}(-\lambda_i)}{|\lambda_i|^2} \}, \text{ and } \frac{1}{2} < \theta, \qquad (3.4)$$

where $\lambda_i \in \text{spectrum}(A_h) \subset \mathbb{C}^-$. We note that method (3.3) cannot be stable if A_h has eigenvalues with positive real part.

For $\theta = 1$ the Euler forward method is obtained. This method is only conditionally stable and it follows from (3.4) that in cases where $\max|\lambda_i|$ is large (stiff problems), the time stepsize k must be chosen relatively small. For the PDE problem (3.1) one thus finds that for stability k must be $O(h^2)$ which is too small for practical purpose. An advantage of the choice $\theta = 1$ is that only matrix-vector multiplications are required - the method is explicit. However, in most applications this advantage is set off by the necessity of taking too many steps resulting in a too large computer time. We also observe that for $\theta = 1$ the order of approximation in time is only $O(k)$ which again suggests the relation $k = O(h^2)$, in order to match the two sources of errors.

For $0 \leq \theta \leq 1/2$ the method is seen to be unconditionally stable - no restriction on k. For $\theta = 0$ one obtains the Euler backward method (Laasonen method) and for $\theta = 1/2$ the trapezoidal rule (Crank-Nicolson). For $\theta = 0$ the order of approximation in time is only $O(k)$ whereas for $\theta = 1/2$ we have $O(k^2)$. However, if $\theta = 1/2$ short wave length perturbations are damped very slowly, which is unfavourable in various practical applications. In this connection, a better choice is $\theta = \frac{1}{2}(1 - k\zeta)$, $\zeta \in \mathbb{R}$ satisfying $h^{-1} \geq \zeta > 0$. This value of θ yields both an $O(k^2)$ time error and sufficient damping of all perturbations. For example, for the problem $u_t = \Delta u$ this choice yields a damping with a typical factor at least equal to

$$e^{-(\zeta + \frac{1}{2d}(\frac{h}{k})^2)t}, \quad t \to \infty,$$

which means that with a large enough parameter ζ sufficient damping is obtained for any choice of k.

Dahlquist has proved that for the popular linear multistep method an accuracy of at most $O(k^2)$ is possible if one requires stability on the whole of $\mathbb{C}^-$. There do exist integration methods - implicit Runge-Kutta methods - for which the error order can be chosen arbitrarily large while maintaining sufficient damping and stability on the whole of $\mathbb{C}^-$ (see [5]). However, such methods require the solution of much more complicated systems of linear algebraic equations than method (3.3) which make them less practical for semi-discrete problems such as (3.2).

For nonlinear systems of the type

$$\frac{dy}{dt} = f(y,t), \quad t>0, \quad y \in \mathbb{R}^N,$$

with f a dissipative operator, i.e., for an inner product $(.,.)$ in $\mathbb{R}^N$ and $\alpha \in \mathbb{R}$ nonpositive, f satisfies

$$(f(y,t) - f(\tilde{y},t), y - \tilde{y}) \leq \alpha(y - \tilde{y}, y - \tilde{y}), \quad \forall t>0, \forall y, \tilde{y} \in \mathbb{R}^N,$$

stability results can be proved for many of the aforementioned methods (see e.g. DEKKER, VERWER [18], VERWER, SANZ-SERNA [50] and, for the θ-method (3.3), AXELSSON [6]).

Runge-Kutta methods, implicit as well as explicit ones, are known to suffer from an order reduction phenomenon. This implies that in applications - semi-discrete PDEs and stiff problems - the true order may be significantly lower than the highest possible order. For stiff problems this was first perceived by PROTHERO and ROBINSON [41]. For semi-discrete PDEs, see VERWER [49] and SANZ-SERNA, VERWER, HUNDSDORFER [44].

For the implicit midpoint rule (or the θ-method with $\theta=1/2$) one can see that this order reduction emanates from the fact that the damping factor can be almost 1. For (almost) constant stepsizes a cancellation effect renders the global error to remain $O(k^2)$, but for variable operators A_h and stepsizes k this cancellation does not take place resulting in an $O(k)$ error (see e.g. AXELSSON [7]). Using a different method of proof, SPIJKER [46] has derived the $O(k)$ result in the maximum norm. For the l_2-norm one can prove (AXELSSON [7]) that a somewhat higher order is obtained if θ is defined in a certain special way as a function of k. SANZ-SERNA, VERWER, HUNDSDORFER [44] have examined other techniques to reduce the order reduction.

In the numerical solution of hyperbolic (wave) equations one encounters other important aspects. A central role is played by the need to work with conservative difference schemes with as little dispersion as possible (VAN DER HOUWEN, SOMMEIJER [34]). A conservative difference scheme conserves one or more relevant energy functionals which, from the physical point of view, is a natural requirement. Any difference scheme suffers from dispersion. Dispersion is the property that the wavelengths of the numerical approximation differ from the exact wavelengths. The effect of too much dispersion is that the solution profile may become wrong, especially so on longer time intervals. This justifies the development of schemes which minimize dispersion.

3.2. Global integration methods

Step-by-step methods advance the solution from t to $t+k$ by relating only approximation values at the time levels $t+k$, t and, possibly, previous time levels. Global methods relate approximate values at all time grid points in the desired interval, $[0, T]$ say, simultaneously. Hence in such a method the time variable is treated like a space variable. Although step-by-step methods are essentially easier to apply, the use of a global method may be beneficial in certain cases, for example, if one wants a stable method of a high order (compare the earlier mentioned $O(k^2)$ barrier of Dahlquist). It is also possible to interpret a high order implicit Runge-Kutta method with many stages as a global time method. However, a more feasible approach is to employ finite difference or finite element methods in time. The problem (3.2) is then considered as a two-point boundary value problem with $U_{h,i}(0) = U_0(\vec{x}_i)$ as a boundary condition at $t=0$ while the differential equation itself is used to define a boundary condition at $t=T$. Details can be found in AXELSSON, VERWER [8]. We observe that for this type of problems a connection exists between certain implicit Runge-Kutta methods and collocation and finite element Galerkin methods. It is also of interest to observe that, as opposed to step-by-step methods, global methods are capable of solving numerically initial value problems which are unstable with respect to perturbations of the initial value.

In the numerical solution of PDEs it may sometimes be desirable to use different time stepsizes in different parts of the spatial domain. This is the case for problems where only in a small part of the domain the solution rapidly varies, whereas in the rest of the domain the solution is nearly stationary. A typical example is a sharp moving front or shock wave. For such problems it is very natural to apply finite element methods based on space-time finite elements. Such a method makes it possible to refine simultaneously in space and time in regions with sharp transitions and to use a coarse space-time mesh in regions with much less variation. This may lead to a considerable reduction of the number of grid points in the time-space domain $\Omega \times [0, T]$ when compared with the standard method of lines approach of using one stepsize k, possibly very small, for all points. The efficient application of such global methods will heavily lean upon the availability of fast iterative solvers for the arising large systems of linear (and nonlinear) algebraic equations. Such a solver has been applied in AXELSSON, STEVENS [9] for the singularly perturbed problem

$$\frac{\partial u}{\partial t} = \epsilon \Delta u + \vec{v} \cdot \nabla + f.$$

3.3. The solution of algebraic systems in implicit methods

We have seen that implicit step-by-step methods require, in each time step, the solution of an algebraic system. As an example, consider system (3.3) arising in the application of the θ-method. This system, with matrix of coefficients $I - (1-\theta)kA_h$ has to be solved in each step. Of coarse, direct methods can be applied, however, as explained in the Introduction, this may be rather time consuming and may require an extensive amount of storage, even in the case of a constant matrix of coefficients (which needs only to be factorized once). Therefore, we shall concentrate on iterative methods.

One possibility is the application of multigrid methods as described in Chapter 2. (This type of method was used in VAN DER HOUWEN, DE VRIES [33] for solving nonlinear PDEs). We remark that, in this case, the condition number of the matrix is $O(1+kh^{-2})$ as $h\to 0$ (with, e.g., $k=O(h)$), and is, therefore, considerably less than in the case of elliptic problems.

Because of the smaller condition numbers, preconditioned iterative methods form an attractive alternative and may perform quite well. For a certain modified preconditioning method the condition number is of order $O(h^{-1/2})$, where $k=O(h)$, which implies that for symmetric systems the number of iterations in, for example, a conjugate gradients method, will increase very slowly as $h\to 0$. Consequently, the method has almost optimal complexity.

Another attractive possibility is the use of 'fractional step' methods. In each time step, such methods solve the system using only a 'fraction' of the operator. From a computational point of view, it is effective to choose the fractions of the operator equal to one-dimensional difference operators. Other possibilities which improve the well known ADI method may be found in VAN DER HOUWEN [35] and MARCHUK, KUZIN [59].

In the following we shall compare the asymptotic computational effort of three different methods (i.e., Euler's explicit method, and the implicit θ-method employing a direct and an iterative process for solving the algebraic system) for integrating the equation

$$\frac{\partial u}{\partial t} = \Delta u, \ \vec{x} \in \Omega \subset \mathbf{R}^d, \ 0 \leq t \leq 1$$

For an explicit method, e.g. Euler's method, we have the stability condition $k \leq \frac{1}{2}d^{-1}h^2$ (cf. (3.4)). Hence, the number of steps is $k^{-1}=2dh^{-2}$. Let the number of elementary algebraic operations (in the matrix-vector multiplications and in the vector sum) be cN, for some constant c, with N denoting the number of unknowns. Then the costs are given by

$$\text{Costs (Expl.)} \sim 2cdh^{-2}N \text{ as } h\to 0.$$

In the implicit θ-method with $\theta=1/2$, we can choose the time steps k arbitrarily as far as stability is concerned, however, for the sake of accuracy we will choose $k=0(h)$. This implies that the space and time discretization error are of the same order. Let the half bandwidth of the matrix A_h be ω, then, when using a direct method, the costs of the factorization for a constant matrix are $\frac{1}{2}\omega^2 N$ and the costs per step for solving the system are $2\omega N$. Since, in most cases, $\omega = h^{-d+1}$, we find for k^{-1} steps

$$\text{Costs (Impl., Dir)} \sim (\tfrac{1}{2}\omega^2 + k^{-1}2\omega)N + k^{-1}cN$$

$$= (\tfrac{1}{2}h^{-d+2} + 2)h^{-d}N + ch^{-1}N \text{ as } h\to 0.$$

Finally, employing as iterative methods in the θ-method we find

Costs (Impl., Iter.)$\sim k^{-1}O(h^{-\nu})N = O(h^{-(1+\nu)})N$ as $h \to 0$,

where the number of iterations is $O(h^{-\nu})$ with $\nu=1/2$ or $\nu=1/4$, depending on the choice of the preconditioning matrix.

Thus, only in one-dimensional problems ($d=1$) the direct method is recommendable, but for $d=2$ and $d=3$ the iterative method is preferable. Also, notice that the explicit method is preferable to the direct method if $d=3$.

3.4. Methods with extended stability intervals

In order to improve the explicit methods various authors have tried to extend the interval of stability of an explicit method. In VAN DER HOUWEN, SOMMEIJER [36] this has been done by combining a second-order, explicit Runge-Kutta method with a stability polynomial with analytically given coefficients. The resulting method allows an easy implementation for an arbitrary number of stages.

Consider the special first-order, m-stage Runge-Kutta method for the initial-value problem

$$\frac{dx}{dt} = f(t,x), \quad t>0, \quad x(t_0) = x_0$$

is of the form

$$x_1 = x_0 + \tau k_m,$$

where

$$k_1 = f(t_0, x_0), \quad k_i = f(t_0 + a_i\tau, x_0 + a_i\tau k_{i-1}), \quad i = 2, \ldots, m.$$

Here, τ is the time step and x_1 an approximation to $x(t_0+\tau)$. Let us apply this method to the model equation $dx/dt = \lambda x$, then $x_1 = P_m(\tau\lambda)x_0$, where $P_m(z) = \sum_{j=0}^{m} \alpha_j z^j$ is the stability polynomial. For first-order methods we require that $\alpha_0 = 1$, $\alpha_1 = 1$. The remaining coefficients can be chosen such that $|P_m(z)| \le 1$, $z \in [-\beta, 0]$ where $\beta = \beta(m)$ and $[-\beta/\lambda, 0]$ is the *stability interval*. Notice that the method parameters $\{a_i\}$ are related to the coefficients $\{\alpha_i\}$ by $a_i = \alpha_i/\alpha_{i-1}$, $i = 2, \ldots, m$.

In order to maximize the length of the stability interval, we choose the shifted Chebyshev polynomial

$$P_m(z) = T_m(\omega_0 + \omega_1 z)$$

with $T_m(\omega_0) = 1$ and $\omega_1 T'_m(\omega_0) = 1$. From this it follows that $\beta = 2m^2$.

For second-order methods we require $\alpha_0 = \alpha_1 = 1$, $\alpha_2 = 1/2$ and the method parameters can be computed as recommended in [36]. This results in $\beta = 2m^2/3$ (this value is already close to the maximal attainable stability limit $\beta_{max} = .82m^2$; in [36] closed-form expressions can be found of stability polynomials with $\beta = .81m^2$).

The above considerations indicate that the stability interval increases with m as $O(m^2)$ whereas the computational costs are $O(m)$ as $m \to \infty$. For the problem $\partial u/\partial t = \Delta u$ considered in the preceding section this implies that

$\tau \leqslant \frac{2}{3} m^2 / \max \lambda(-A_h) = \frac{2}{3} m^2 / 2dh^{-2} = (mh)^2 / 3d$. Hence, choosing $m = \mu h^{-\frac{1}{2}}$, with μ a given constant, we obtain the stability condition $\tau \leqslant \mu^2 h / 3d$, i.e. $\tau = O(h)$. This is just the situation for implicit methods. It should be noted that the 'extended stability' methods are related to the Chebyshev semi-iterative method for solving the equations occurring in implicit step-by-step methods.

The computational costs of the 'extended stability' methods are given by

$$\text{Costs (Expl., Ext.)} = O(h^{-1.5})N.$$

A comparison with implicit methods using special preconditioned iterative methods reveals that the 'extended stability' methods are still more expensive; however, they have the advantage that, when applied to nonlinear problems, there is no need to evaluate the Jacobian matrix as is the case in most iterative methods. On the other hand, in order to satisfy the stability condition of the Runge-Kutta method we need an estimate of the spectral radius of the Jacobian. A well known technique for estimating the spectral radius uses the power method and successive differences of the right-hand side function $f(t,x)$: let v be an approximation to the eigenvector of Jacobian matrix J with eigenvalue λ equal to the spectral radius. Then

$$\epsilon^{-1}[f(t,x+\epsilon v) - f(t,x)] \approx J(t,v)v \approx \lambda v.$$

In [52] satisfactory results are reported for this type of method.

The literature on the numerical solution of time-dependent PDEs is considerable. A few not yet mentioned reports of DOUGLAS, DUPONT, EWING, ZLAMAL, HELFRICH, LUSKIN and of the present author may be found in the references of [10]. A survey paper of higher-order methods has recently been published by SEWARD et al. [45]. References to papers on the numerical solution of convection-diffusion equations of HUGHES, JOHNSON and others may be found in [9].

As to the numerical solution of the Euler equations in flow problems, of Stokes equations, and of Navier-Stokes equations, the reader is referred to the papers by, e.g., HUGHES et al. and RAVIART et al. For transonic flow problems see, e.g., the papers of GLOWINSKI, PERIAUX et al., VIVIANT, JAMESON, BOERSTOEL, RIZZI, etc.

4. SOME REMARKS ON THE USE OF SUPERCOMPUTERS IN THE NUMERICAL SOLUTION OF PARTIAL DIFFERENTIAL EQUATIONS

For the solution of the large systems of difference equations which arise in the numerical solution of partial differential equations in three dimensions, for example, it is necessary to restrict the use of memory as much as possible. Otherwise, too much time and computer costs will be spent to the transport (I/O) of data between the (fast) central memory and the (slow) peripheral memory.

Hence, the sparseness of the difference matrix should be maintained, so that iterative methods are to be preferred above direct methods. Furthermore, also the use of supercomputers with very fast and, sometimes, parallel arithmetic may be necessary.

There are a number of survey papers in this area, like HELLER [60] and ORTEGA and VOIGT [55]. The use of supercomputers in computational mathematics is described by TE RIELE [56]. For the use of the BLAS-routines (Basic Linear Algebra Subprograms) on the CYBER 205, cf. LOUTER-NOOL [58]. An introductory survey paper about the use of supercomputers in the multigrid method was written recently by HEMKER [29]. For incomplete factorization methods on supercomputers, cf. VAN DER VORST and VAN KATS [57], and for blockmatrices, cf. [4].

A portable vector-code for multigrid modules has appeared recently in HEMKER, WESSELING and DE ZEEUW [26] and in HEMKER, DE ZEEUW [30].

The subject linear algebra and supercomputers has become a large research area, as appears for instance from a number of talks which were presented recently at a supercomputer conference in Norfolk, Va, USA, in November 1985.

As a conclusion from the above, it is clear that the research in the field of partial differential equations, as carried out by the Department of Numerical Mathematics of the CWI, covers a broad field and has reached very important results.

REFERENCES
1. G.P. ASTRACHANCEV (1971). An iterative method of solving elliptic net problems. *Z. vyčisl. Mat. mat. Fiz. 11,* 2, 439-448.
2. A.K. AZIZ (ed.) (1972). *The Mathematical Foundations of the Finite Element Method with Applications to Partial Differential Equations,* Academic Press, New York.
3. O. AXELSSON, V.A. BARKER (1984). *Finite Element Solution of Boundary Value Problems. Theory and Computation,* Academic Press, Orlando.
4. O. AXELSSON. Analysis of incomplete matrix factorizations as multigrid smoothers for vector and parallel computers. *Applied Mathematics and Computation,* to appear.
5. O. AXELSSON (1969). A class of *A*-stable methods. *BIT 9,* 185-199.
6. O. AXELSSON (1984). Error estimates over infinite intervals of some discretizations of evolution equations. *BIT 24,* 413-424.
7. O. AXELSSON. Stability and error estimates valid for infinite time, for strongly monotone and infinitely stiff evolution equations. To appear in *Proceedings EQUADIFF VI,* Brno, 1985.
8. O. AXELSSON, J.G. VERWER (1985). Boundary value techniques for initial value problems in ordinary differential equations. *Math. Comp. 45,* 153-171.
9. O. AXELSSON, S. STEVENS. *A Novel Defect-Correction Method for Convection-Diffusion Problems,* preprint, Department of Mathematics, Catholic University, Nijmegen.
10. O. AXELSSON (1980). Computational aspects in the numerical solution of parabolic problems by finite element methods. J.G. VERWER (ed.).

Colloquium Numerical Solution of Partial Differential Equations, Stichting Mathematisch Centrum, MC Syllabus 44, pp. 1-9, Amsterdam.

11. N.S. BACHVALOV (1966). On the convergence of a relaxation method with natural constraints on the elliptic operator. *Z. vyčisl. Mat. mat. Fiz. 6, 5*, 861-883.
12. R.E. BANK, T. DUPONT (1981). An optimal order process for solving finite element equations. *Math. Comp. 36*, 35-51.
13. E.B. BECKER, G.F. CAREY, J.T. ODEN (1981, 1983, 1984). Finite Elements, Vols. I, II, III, Prentice-Hall, Englewood Cliffs, N.J.
14. D. BRAESS (1981). The contraction number of a multigrid method for solving the Poisson equation. *Numer. Math. 37*, 387-404.
15. A. BRANDT. Guide to multigrid development, in [21], 220-312.
16. P.G. CIARLET (1978). *The Finite Element Method for Elliptic Problems*. North-Holland Publ., Amsterdam.
17. S.F. MCCORMICK, J.W. RUGE (1982). Multigrid methods for variational problems. *SIAM J. Numer. Anal. 19*, 924-929.
18. K. DEKKER, J.G. VERWER (1982). *Stability of Runge-Kutta Methods for Stiff Nonlinear Differential Equations*, North-Holland, Amsterdam.
19. J.E. DENDY, JR. (1982). Black box multigrid. *J. Comput. Phys. 48*, 366-386.
20. R.P. FEDORENKO (1961). A relaxation method for solving elliptic difference equations. *Z. vyčisl. Mat. mat. Fiz. 1, 5*, 922-927.
21. W. HACKBUSH, U. TROTTENBERG (eds.) (1982). *Multigrid Methods, LNM Vol. 960*, Springer-Verlag, Berlin.
22. W. HACKBUSH (1980). Survey of convergence proofs for multigrid iteration. J. FREHSE ET AL. (eds.). *Special Topics in Applied Mathematics*, 151-164, North-Holland Publishing Company, Amsterdam.
23. P.W. HEMKER. *The Use of Defect Correction for the Solution of a Singularly Perturbed ODE*, preprint.
24. P.W. HEMKER (1984). Mixed defect correction for the solution of a singular perturbation problem. *Computing, Suppl. 5*, 123-145.
25. P.W. HEMKER (1980). Introduction to multigrid methods. J.G. VERWER (ed.). *Colloquium Numerical Solution of Partial Differential Equations*, 59-97, Mathematisch Centrum, Amsterdam.
26. P.W. HEMKER, R. KETTLER, P. WESSELING, P.M. DE ZEEUW (1983). Multigrid methods: development of fast solvers. *J. Appl. Math. and Comp. 13*, 311-326.
27. P.W. HEMKER, S.P. SPEKREIJSE (1985). Multigrid solution of the steady Euler equations. D. BRAESS ET AL. (eds.). *Advances in Multigrid Methods*, 33-44, Vieweg, Braunschweig/Wiesbaden.
28. P.W. HEMKER, S.P. SPEKREIJSE (1985). *Multiple Grid and Osher's Scheme for the Efficient Solution of the Steady Euler Equations*, Report NM-R8507, CWI, Amsterdam.
29. P.W. HEMKER (1984).Multigrid algorithms run on supercomputers. *CWI Newsletter 3*, 25-30.

30. P.W. HEMKER, P.M. DE ZEEUW (1985). Some implementations of multigrid linear system solvers. D.J. PADDON, H. HOLSTEIN (eds.). *Multigrid Methods for Integral and Differential Equations*, 85-116, Clarendon Press, Oxford.
31. A.J. HOFFMAN, M.S. MARTIN, D.J. ROSE (1973). Complexity bounds for regular finite difference and finite element grids. *SIAM J. Numer. Anal. 10*, 364-369.
32. P.J. VAN DER HOUWEN, B.P. SOMMEIJER (1981). *Analysis of Richardson Iteration in Multigrid Methods for Nonlinear Parabolic Differential Equations*, Report NW 105/81, Mathematisch Centrum, Amsterdam.
33. P.J. VAN DER HOUWEN, H.B. DE VRIES (1982). Preconditioning and coarse grid corrections in the solution of the initial value problem for nonlinear partial differential equations. *SIAM J. Sci. Stat. Comput. 3*, 473-485.
34. P.J. VAN DER HOUWEN, B.P. SOMMEIJER (1985). *Explicit Runge-Kutta (-Nyström) Methods with Reduced Phase Errors for Computing Oscillating Solutions*, Report NM-R8504, CWI, Amsterdam.
35. P.J. VAN DER HOUWEN (1984). Iterated splitting methods of high order for time-dependent partial differential equations. *SIAM J. Numer. Anal. 21*, 635-656.
36. P.J. VAN DER HOUWEN, B.P. SOMMEIJER (1980). On the internal stability of explicit m-stage Runge-Kutta methods for large m-values. *ZAMM 60*, 479-485.
37. R. KETTLER, P. WESSELING. Aspects of multigrid methods for problems in three dimensions. *Applied Mathematics and Computation*, to appear.
38. A.R. MITCHELL (1969). *Computational Methods in Partial Differential Equations*, Wiley, London.
39. A.R. MITCHELL, R. WAIT (1977). *The Finite Element Method in Partial Differential Equations*, Wiley, London.
40. R.A. NICOLAIDES (1977). On the l^2 convergence of an algorithm for solving finite element equations. *Math. Comp. 31*, 892-906.
41. A. PROTHERO, A. ROBINSON (1974). The stability and accuracy of one-step methods. *Math. Comp. 28*, 145-162.
42. LIN QUN, ZHU QIDING (1984). Asymptotic expansions for the derivative of finite elements. *J. Comp. Math 2*.
43. R.D. RICHTMYER, K.W. MORTON (1967). *Difference Methods for Initial-Value Problems*, Interscience, New York.
44. J.M. SANZ-SERNA, J.G. VERWER, W.H. HUNDSDORFER (1985). *Convergence and Order Reduction of Runge-Kutta Schemes Applied to Evolutionary Problems in Partial Differential Equations*, Report NM-R8525, CWI, Amsterdam.
45. W.L. SEWARD, G. FAIRWEATHER, R.L. JOHNSTON (1984). A survey of higher-order methods for the numerical integration of semidiscrete parabolic problems. *IMA J. Numer. Anal. 4*, 375-425.
46. M.N. SPIJKER. Stepsize restrictions for stability of one-step methods in the numerical solution of initial value problems. *Math. Comp.*, to appear.

47. G. STRANG, G.J. FIX (1973). *An Analysis of the Finite Element Method,* Prentice-Hall, Englewood Cliffs, N.J.
48. V. THOMÉE (1984). *Galerkin Finite Element Methods for Parabolic Problems, LNM 1054,* Springer-Verlag, Berlin.
49. J.G. VERWER. Convergence and order reduction of diagonally implicit Runge-Kutta schemes in the method of lines. *Proceedings of the Numerical Analysis Conference Dundee 1985,* to appear.
50. J.G. VERWER, J.M. SANZ-SERNA (1984). Convergence of method of lines approximations to partial differential equations. *Computing 33,* 297-313.
51. P. WESSELING (1982). Theoretical and practical aspects of a multigrid method. *SIAM J. Sci. Stat. Comput. 3,* 387-407.
52. A. WOLFBRANDT, K.-E. KARLSSON (1984). *A New Explicit Technique of Calculating the Electromagnetic Field and Power Losses in Ferromagnetic Materials,* Preprint, ASEA, Västerås, Sweden.
53. H. YSERENTANT (1983). *On the Multilevel Splitting of Finite Element Spaces,* Bericht Nr. 21, 1983, Institut für Geometrie und Praktische Mathematik, RWTH Aachen.
54. P.M. DE ZEEUW, A.J. VAN ASSELT (1985). The convergence rate of multilevel algorithms applied to the convection-diffusion equation. *SIAM J. Sci. Stat. Comput. 6,* 492-503.
55. J.M. ORTEGA, R.G. VOIGT (1985). Solution of partial differential equations on vector and parallel computers. *SIAM Review 27,* 149-240.
56. H.J.J. TE RIELE (1985). *Applications of Supercomputers in Mathematics,* Report NM-N8502, CWI, Amsterdam.
57. H.A. VAN DER VORST, J.M. VAN KATS (1984). *The Performance of some Linear Algebra Algorithms in FORTRAN on CRAY-1 and CYBER 205 Supercomputers,* Techn. Report TR-17, ACCU, Utrecht.
58. M. LOUTER-NOOL (1985). BLAS on the CYBER 205, Report NM-R8524, CWI, Amsterdam.
59. G.I. MARCHUK, V.I. KUZIN (1983). On the combination of finite element and splitting-up methods in the solution of parabolic equations. *J. Comp. Physics 52,* 237-272.
60. D. HELLER (1978). A survey of parallel algorithms in numerical algebra. *SIAM Review 20,* 740-777.

Dynamics in Bio-Mathematical Perspective

Odo Diekmann
Centre for Mathematics and Computer Science
P.O. Box 4079, 1009 AB Amsterdam, The Netherlands
and
Institute of Theoretical Biology, University of Leiden
Groenhovenstraat 5, 2311 BT Leiden, The Netherlands

1. INTRODUCTION

1.1. Biology and mathematics: a continual interaction
In mathematics and science it is nowadays almost compulsory to follow the narrowing road of specialization. In a period in which the would-be universal scientist is forced to read night and day (and even while doing so is confronted with an ever increasing back-log), intense co-operation between specialists in different fields seems to be a designated way to escape from the various pitfalls (the Scylla of narrowmindedness and the Charybdis of unproductivity). This negative argument in favour of interdisciplinary projects is easily supplemented with more positive ones, such as: co-operation between people having different backgrounds increases the chances of discovering unexpected but enlightening connections and, last but not least, may enhance working pleasure considerably.

The interplay of mathematics and the sciences is not an instantaneous one-way process but rather a process of repeated cross-fertilization. Foggy notions and questions about real world phenomena have to be clarified when one tries to reformulate them in terms of a mathematical model. The incorporation of specific models (and the problems they pose) within a mathematical framework of some generality serves as a test for the mathematical structure itself and may lead to the creation of a new, extended and improved structure based on a deeper understanding. The outcome of a mathematical analysis may trigger renewed investigations, with different eyes, of the natural phenomena which one is trying to describe and understand.

In this lecture I intend to illustrate the general statements above by means of a few selected examples. These examples have in common that they are concerned with dynamics, the time-evolution of states, in the context of biological (more precisely, population dynamical and epidemiological) models. This characteristic provides a first justification and interpretation of the title. A second interpretation derives from the fact that the interaction between biology and mathematics is itself a dynamical process. I will try to describe the examples in such a way that at least part of this process becomes visible. I will stress the mutual influence by paying special attention to the way things have developed to what they are now (and by speculating a little bit about future developments). Of course there are many cases in which by now well-known mathematical techniques are used to answer by now well-defined biological questions but, however useful that may be, this is not the kind of applications of mathematics in biology I want to describe. Instead I will concentrate on situations in which the mathematical and the biological aspects coevolve towards a state in which they are adapted to each other at the benefit of both. Inevitably the composition of the audience and my own background create some bias to the effect that the mathematical aspects will be overemphasized.

Many interesting and important recent results and developments of dynamical systems theory are not touched upon in this lecture (no chaos, for instance). Most of the work (even of that with a biological flavour) in which the Department of Applied Mathematics of the Centre for Mathematics and Computer Science (and its predecessor, the Mathematical Centre) was involved during the last 40 years, will not be described. I concentrate on two problems which, I feel, are well suited to illustrate some general features of the coevolution of mathematics and science, which are more or less representative of the work done at the Department of Applied Mathematics, and which are interesting by themselves. The solution of the first problem requires hard nonlinear analysis (up to six or seven constants have to be chosen suitably to get the estimates right). The solution of the second problem is based on soft linear functional analysis (an abstract framework has to be defined to make things easy and straightforward).

Chapter 2 deals with the first problem, the description and analysis of the geographical spread of an infectious disease. In Section 1.2 I give a preview of the main questions and answers while emphasizing the conceptual aspects and neglecting the technical ones.

Finding an appropriate mathematical framework for models of physiologically structured populations is the main issue of Chapter 3. Although biologically not the most interesting case, I concentrate on age structured populations for didactic reasons (to understand the equations of age dependent population dynamics requires comparatively little energy of the uninitiated reader; see [64,51] for a systematic exposition of models and equations in the general case and for a snapshot of the state-of-the-art of the rapidly growing mathematical theory). An introductory preview of the basic ideas and problems is given in Section 1.3.

1.2. The speed of propagation and intermediate asymptotics

In Chapter 2 we consider a situation with very simple dynamics. A steady state, called 0, is unstable and any biologically realizable perturbation, no matter how small, gives rise to a sequence of events (an orbit) which ends in a stable steady state, called ∞. Real world examples range from fires (combustion theory), over the development of an infectious disease to the taking over by a favourable mutant gene. Despite the dynamical simplicity one can ask a difficult question: how fast will the transition $0 \to \infty$ effectively take place. The sting is in the adverb 'effectively', which makes the answer 'It will take an infinite time' inappropriate. The mathematical theory of dynamical systems centers around the asymptotic behaviour of trajectories for large time and, in particular, the classification of limit sets. Transients are the Cinderellas which do the hard and dirty work, but which are hardly ever regarded as interesting by themselves.

Our question can be rephrased in terms of the physical notion of 'time scale' (see, for instance, LIN, SEGEL [47]),but in a nonlinear problem several time scales can be involved (in the present case one has at least three phases: an initial phase governed by the linearization near 0, an intermediate phase governed by the nonlinearity and a final phase governed by the linearization near ∞). So do we have to take recourse to numerical calculations, taking for granted the inherent imperfection that variation of parameters may lead to large amounts of numbers from which it is hard to deduce the essential information?

Let us first indulge in our basic question, while concentrating, for the sake of exposition, on the case of an infectious disease affecting some agricultural crop. A farmer finding his wheat-field invaded by a certain rust wants to estimate how much of the field will be unaffected at harvest time (note that the upper limit for the time window accentuates that the problem does not fit into the standard large time asymptotic realm). It appears that the problem has a spatial dimension too. At first sight this only seems to complicate the matter but, as we will see, it actually enables us to bring asymptotics back into the play.

Assume, as an 'idealization', that the field extends infinitely far in all directions. Then we can look for travelling plane waves, a special kind of self-similar solutions. The rationale for our interest in these special solutions lies in the idea that an observer moving with the right speed might be able to study the transients. Or, in other words, in a moving coordinate system the transients may look like 'frozen' spatial transitions.

A robust conclusion obtains: travelling plane waves exist for all speeds $c \geq c_0$ for some c_0 and this minimal wave speed c_0 is the asymptotic speed of propagation of disturbances in a sense which is on the one hand excellently adapted to the biological connotation and, on the other hand, mathematically precise. By 'robust' we mean that the conclusion is valid for a large class of models which are quite different from a mathematical point of view, yet describe biologically similar phenomena. The equations corresponding to these models take divergent forms as is manifest from the adjectives: reaction-diffusion, integro-differential, integro-difference, Volterra-Hammerstein.

Comparison theorems and the construction of suitable lower- and upper-solutions are indispensable tools for their analysis.

It is an experimental fact, derived from simulation studies, that the quantity c_0 is highly relevant for a description of propagation in *finite* fields during *finite* time intervals. In the interesting book *Similarity, Self-Similarity and Intermediate Asymptotics* [7] G.I. BARENBLATT writes:

> "Self-similar solutions also describe the 'intermediate asymptotic' behaviour of solutions of wider classes of problems in the range where these solutions no longer depend on the details of the initial and/or boundary conditions, yet the system is still far from being in a limiting state"

(and he stresses the importance of self-similar solutions as an aid in interpreting large amounts of data obtained from computer simulations). Unfortunately it appears to be rather hard to prove (or even formulate) precise mathematical statements about intermediate asymptotics (and I cannot resist the temptation of writing a commonplace: this subject deserves to be more widely and deeply studied!). However, even though the theoretical basis is perhaps not as solid as it should be, we arrive at a clear-cut conclusion: the transition $0 \to \infty$ takes place with a well-defined speed c_0.

Once such a strong result is available, it becomes worth-while to embark upon a more detailed modelling exercise dealing with such questions as: how do the ingredients of the model relate to measurable biological quantities? Moreover, the computation of c_0 from the ingredients is a point of concern and, finally, the prediction of c_0 found from the model should be tested against the speed found in the field (measurements usually indeed display a constant rate of expansion!).

1.3. About states and state-spaces

In order to give a realistic description of disease propagation it does not suffice to classify an individual plant as either healthy or infected. The production of infectious agents (say spores) is determined by the state of the particular plant, where 'state' should incorporate everything relevant for determining the spore production now and in the future, given the course of the environment (the weather, for instance). This is not an unusual situation. Individuals are not really the 'atoms' of population dynamics, simply because they differ in traits as age, size, energy reserves etc., which are of great influence on their population dynamical behaviour (giving birth, dying, consumption of limiting nutrients, occupying territoria etc.). An obvious idea is to introduce a (finite dimensional) *individual state space* Ω and to conceive of the population as a frequency distribution (sometimes called the population density) n over Ω. The dynamics of the individuals (their ageing, growing, metabolism, etc.) are described by ordinary differential equations and simple bookkeeping arguments at the population level lead to a first order partial differential equation for n. These partial differential equations may exhibit several unusual features:

birth terms are non-local and the support of n may concentrate on a lower dimensional manifold in Ω.

A convenient conceptual framework for the description of dynamical phenomena can be build from the notions of *state, next-state operators* and *generator* (and, in addition, *input* and *output* but these are not essential for our purposes now). In the present context the notion of state figures at two levels. At the individual level the state corresponds to the finitely many characteristics, say summarized in a vector x, which uniquely fix the population dynamical 'status' of an individual. The variable x takes values in Ω, a subset of $\mathbb{R}^k$. At the population level the state is given by the frequency distribution n and we have still to specify to which space X of functions on Ω $n(t)$ is assumed to belong.

Operators $T(t,t_0)$ map the population state at t_0 onto the population state at time t, thereby providing a complete description of the dynamics. Even though the collection of operators $T(t,t_0)$ is just a mathematical incarnation of its real world counterpart it is usually impossible to give a direct mathematical definition. They have a clear and well-defined interpretation but, as a rule, it is impossible to calculate explicitly how they act on the basis of nothing but modelling assumptions. Instead we usually first derive the (infinitesimal) *generator* $A(t_0)$ by calculating changes of the state in small time intervals h up to first order in h and, after dividing by h, taking the limit $h\downarrow 0$. Hence $A(t_0)$ is, at least formally, the derivative of $T(t,t_0)$ with respect to t evaluated at $t=t_0$. The advantage of the 'infinitesimal' formulation is that the different contributions to the dynamics from the various 'forces' are uncoupled in the limit $h\downarrow 0$ whereas, in contrast, they are strongly intermingled in finite time intervals (an individual which has died cannot give birth!). The 'local' differential equation $\frac{dn}{dt}=A(t)n$ is much easier derived from a verbal description of a model then the 'global' solution operators $T(t,t_0)$. This is, of course, one of the main reasons for the omnipresence of differential equations in (applied) mathematics.

Part of the bookkeeping arguments alluded to above are formal $h\downarrow 0$ calculations which yield the equation $\frac{dn}{dt}=A(t)n$ in the form of a partial differential equation supplemented with appropriate boundary conditions. So here $A(t)$ is a differential (or integro-differential or differential-difference) operator acting on functions of the variable x. In this derivation we don't bother about the precise definition of the population state space X or about the sense of convergence as $h\downarrow 0$. In the partial differential equation formulation we think of n as a function of two variables, $n(t,x)=n(t)(x)$, and neither X nor the sense in which the equation should hold is specified during a derivation by formal calculus.

Partly for the sake of exposition and partly because more general population models are not elaborated in detail yet, we assume from now on that the environmental circumstances are constant in time. So experiments starting from the same initial state are identical, whether we perform them now or two weeks from now. Time translations don't matter then and, slightly abusing notation, we may write $T(t,t_0) = T(t-t_0)$ and assume that A is independent of t. Moreover, let us assume that density dependence may be neglected such that, as a consequence, all our operators will be linear.

In any book on the functional analytic theory of semigroups (HILLE and PHILLIPS [42], BUTZER and BERENS [14], DAVIES [18], PAZY [56], GOLDSTEIN [33], VAN CASTEREN [15], NAGEL [53]) one finds the following definitions. Let X be a Banach space with norm $\|\cdot\|$, and let for each $t \geq 0$, $T(t)$ be a bounded linear operator on X. Assume that:
(i) $T(0) = I$, where I denotes the identity operator on X,
(ii) $T(t+s) = T(t)T(s)$, $t,s \geq 0$,
(iii) $\lim_{t \downarrow 0} \|T(t)\phi - \phi\| = 0$, for all $\phi \in X$.

Then $\{T(t)\}$ is called a *strongly continuous semigroup* (of bounded linear operators) on X.

The prefix 'semi' reflects the restriction $t \geq 0$. Note that (i) and (ii) yield a mathematical formulation of intuitive ideas about next-state operators. The condition (iii) is, as one can easily verify by exploiting (i) and (ii), equivalent with the condition that orbits are continuous, i.e. for each $\phi \in X$ the map $t \mapsto T(t)\phi$ is continuous from $\mathbb{R}_+$ to X.

The infinitesimal generator A of $\{T(t)\}$ is the, in general unbounded, operator defined by

$$A\phi = \lim_{h \downarrow 0} \frac{1}{h}(T(h)\phi - \phi)$$

whenever the limit exists. So $D(A)$, the domain of A, is by definition the set of $\phi \in X$ for which this limit exists.

Although we use the same symbols and terminology, we are at the moment dealing with two different 'worlds'. In one lives a formally derived partial differential equation, in the other an unspecified semigroup and generator acting on an unspecified Banach space X. It seems conceivable to make the connection by removing the largely conceptual difference between $n(t,x)$, a function of two variables, and $n(t)(x)$, a function of t with values in a space X of functions of x. But is this worth the effort? Does an abstract approach make life easy? A controversial question to which different people may give opposite answers.

One of the high-lights of semigroup theory is the Theorem of Hille and Yosida which gives a precise characterization of the generators of strongly continuous semigroups. So if we make a choice for the function space X and define, on the basis of the appearance of the partial differential equation, the operator A, in particular its domain, we may try to verify the necessary and sufficient conditions of the Hille-Yosida Theorem. If we are successful this yields an existence and uniqueness result for solutions of the time evolution problem. So here we first reinterpret our partial differential equation as an equation of the form $\frac{dn}{dt} = An$, then associate with A the semigroup $T(t)$ and finally define $n(t,x,\phi) = (T(t)\phi)(x)$, where $\phi(x) = n(0,x)$ is the initial condition at $t=0$ which is (assumed to be) given. This is a usual procedure for dealing with parabolic equations, where A is an elliptic operator for which a large body of results about spectrum and resolvent estimates, the key ingredients of a verification of the Hille-Yosida conditions, is available (see HENRY [39] or FRIEDMAN [32]).

When dealing with physiologically structured population models (or with delay equations, i.e. differential equations which do incorporate some influence of the past on the future, see HALE [36]) we proceed differently. The solution $n(t,x,\phi)$ of the initial value problem is rather easily defined constructively (see section 3.1 for an example). Next we define $T(t)\phi=n(t,\cdot,\phi)$ and calculate from this definition the generator A. So here we obtain only a posteriori a reinterpretation of the partial differential equation as the abstract ordinary differential equation $\frac{dn}{dt}=An$ and the profit is far from self-evident.

In the linear case a basic advantage of the semigroup approach derives from available results concerning the connection between the spectrum of A and the asymptotic behaviour of $T(t)$ (some of the more recent results in this area were motivated by models from age dependent population dynamics! See PRÜSS [57-59] and WEBB [69]). In the case of ordinary differential equations in $\mathbb{R}^k$ this is just the connection between the eigenvalues of the matrix A and the asymptotic behaviour of solutions. But in an infinite dimensional situation there may exist spectral values which are not eigenvalues and a careful analysis is needed. I don't review this interesting theory here, but confine myself to remarking that it serves as a mayor motive for putting specific evolution problems in the semigroup framework. The very recent and highly interesting lecture notes *One-Parameter Semigroups of Positive Operators* [53], edited by R. NAGEL, gives a wealth of results culminating in an extensive study of the special (but rather important also from an 'applied' point of view) case of positive operators. Also see HEIJMANS [40,41]. DIEKMANN, METZ, KOOIJMAN and HEIJMANS [25] or WEBB [69] for an exposition directed towards applications in population dynamics.

Bypassing a vast literature on the generation of nonlinear semigroups (e.g. BARBU [6], BREZIS [12], CRANDALL [17]), we recall that in local stability and bifurcation theory one deals with perturbations of linear problems. Many results in this area can be obtained from simple estimates and the implicit function theorem once has formulated the appropriate variant of the *variation-of-constants formula*

$$T(t)=T_0(t)+\int_0^t T_0(t-\tau)BT(\tau)d\tau.$$

Here $T_0(t)$ is a semigroup generated by A_0, B is a bounded perturbation and $T(t)$ is the semigroup generated by A_0+B. In stability and bifurcation problems B is small in an appropriate sense but not necessarily linear. The variation-of-constants formula enables us to estimate how the smallness of B affects the solution operators $T(t)$ and to prove the principle of linearized stability, the center manifold theorem etc. in completely the same way as one does in the case of ordinary differential equations. As a side-remark we mention that an appropriate form of *relative* boundedness of B is sufficient for this purpose (see, for instance, HENRY [39]).

We conclude that a basic advantage of the semigroup approach is that one can prove many results once and for all in the general setting such that subsequently one can draw conclusions about solutions of specific evolution equations by showing that the general results apply.

Following this approach in the case of physiologically structured population models (and in the case of delay equations as well) we run into some disappointment: the general abstract framework does not fit as good as one feels it ought to fit! The problem that arises is explained in Section 3.2 by means of an example. Rather than concluding that the 'basic advantage' is not so big after all and sitting down under it, we take up the challenge, analyse the difficulty and find that the equations do fit excellently within a somewhat extended general framework. In retrospect the extension is quite natural from a mathematical point of view as well and one can easily explain the framework in mathematical terms, without any reference to models from population dynamics or any other application. We emphasize, however, that the tension between general theory and specific applications (as exemplified in feelings of irritation and frustation: why are these damned problems so resistant against an abstract approach which intends to make them easy instead of difficult!?) serves as a catalyser for finding the key ideas.

The work on physiologically structured population models has only just begun and much remains to be done. At the end of the paper I will stress the need for young talented people to carry out the program.

2. THE GEOGRAPHICAL SPREAD OF AN INFECTIOUS DISEASE

2.1. A mathematical prototype: linear diffusion
In this section I will present some rather simple explicit calculations which, I hope, illuminate the main concepts and results. The simplest differential equation

$$\dot{u} = ku \tag{2.1}$$

states that the rate of production of 'particles' (say genes or spores) is proportional, with constant k, to their density u. Assume $k > 0$. Then $u = 0$ is an unstable steady state and, in some sense, $u = \infty$ is a stable steady state. Next suppose our particles are subject to random spatial migration in a plane and replace (2.1) by the diffusion equation

$$\frac{\partial u}{\partial t} = D \Delta u + ku \tag{2.2}$$

where

$$\Delta u = \frac{\partial^2 u}{\partial x_1^2} + \frac{\partial^2 u}{\partial x_2^2}$$

and where the diffusion coefficient D is a measure for the variance of the motion. The fundamental solution

$$u(t,x) = \frac{1}{4\pi Dt} e^{-\frac{|x|^2}{4Dt} + kt} = \frac{1}{4\pi Dt} e^{kt\left(1 - \frac{|x|^2}{4kDt^2}\right)} \tag{2.3}$$

describes what happens when we start at $t=0$ with one particle located at $x=0$. From this explicit expression it follows immediately that for any fixed $\epsilon > 0$

$$u(t,x) \xrightarrow{t \to \infty} \begin{cases} 0 & \text{if } |x|^2 > (4Dk + \epsilon)t^2 \\ \infty & \text{if } |x|^2 < (4Dk - \epsilon)t^2 \end{cases} \tag{2.4}$$

So, asymptotically for $t \to \infty$, nothing has happenend yet outside growing circles of radius $t\sqrt{4Dk + \epsilon}$ and everything has happened already inside growing circles of radius $t\sqrt{4Dk - \epsilon}$. Therefore we call

$$c_0 = 2\sqrt{Dk} \tag{2.5}$$

the *asymptotic speed of propagation* of disturbances (the need to provide c with an index will become evident soon).

Two questions arise:
(i) can we obtain more information about the structure of the transition $0 \to \infty$ in the vicinity of the boundary of the growing circles?
(ii) is it possible to derive (or at least guess) the speed $c_0 = 2\sqrt{Dk}$ a priori, i.e. without solving equation (2.2) explicitly?

It will appear that the answer to (i) provides a first step towards the answer of (ii).

So far we have exploited the radial symmetry of the fundamental solution (2.3) by concentrating at circles, i.e. using $|x|^2$ as our basic variable. But let us now choose some arbitrary unit vector ζ and look explicitly in the direction of ζ by taking for x a representation

$$x = \alpha(t, \theta)\zeta + y, \quad \text{with } y \cdot \zeta = 0, \tag{2.6}$$

where θ represents a 'local' one-dimensional coordinate and the scalar function α has still to be determined. Upon substitution in (2.3) we find

$$u(t,x) = \frac{1}{4\pi Dt} e^{kt\left(1 - \frac{\alpha^2}{4kDt^2}\right)} e^{-\frac{|y|^2}{4Dt}} \tag{2.7}$$

which is bounded away from 0 and ∞ for $t \to \infty$ provided we make sure that

$$\alpha^2(t,\theta) = 4kDt^2\left(1 - \frac{\ln t}{kt} + \frac{h(t,\theta)}{kt}\right)$$

for some bounded function h. For the special choice

$$\alpha^2(t,\theta) = 4kDt^2\left(1 - \frac{\ln t}{kt} - \frac{\ln 4\pi D}{kt} + \frac{\theta}{\sqrt{kDt}}\right) \tag{2.8}$$

we find that for $t \to \infty$

$$u(t,x) \to e^{-\sqrt{\frac{k}{D}}\theta} \tag{2.9}$$

uniformly for y and θ in compact subsets (note that the t-dependent constraint on the range of θ 'dissolves' in the limit $t \to \infty$). The formula (2.8) implies that

$$\alpha(t,\theta) = m(t) + \theta + O(\frac{\ln^2 t}{t}), \quad t \to \infty, \tag{2.10}$$

where

$$m(t) = 2\sqrt{Dk}\, t\, \sqrt{\frac{D}{k}} \ln t - \sqrt{\frac{D}{k}} \ln 4\pi D \tag{2.11}$$

So asymptotically for $t \to \infty$ the solution behaves in the direction ζ like a plane wave (no dependence on y!) of the form $\exp(-\sqrt{\frac{k}{D}}\theta)$ which travels approximately with speed

$$\dot{m}(t) = 2\sqrt{Dk} - \sqrt{\frac{D}{k}}\frac{1}{t}. \tag{2.12}$$

Since ζ is arbitrary we conclude that the solution u 'decomposes' into plane waves travelling in all directions with speed $2\sqrt{Dk}$ and that these waves describe the transition between the inside of the circles ($\theta \to -\infty$) and the outside ($\theta \to +\infty$).

We could as well search for travelling plane wave solutions of the diffusion equation (2.2) directly. Substituting

$$u(t,x) = w(x \cdot \zeta - ct) \tag{2.13}$$

we find for w the ordinary differential equation

$$Dw'' + cw' + kw = 0 \tag{2.14}$$

where a prime denotes the derivative with respect to the variable

$$\theta = x \cdot \zeta - ct. \tag{2.15}$$

The solutions of (2.14) are of the form $w(\theta) = C\exp(\lambda\theta)$ with

$$\lambda = \frac{-c \pm \sqrt{c^2 - 4Dk}}{2D} \tag{2.16}$$

and C an arbitrary constant. The biological interpretation requires that w is non-negative. Consequently we are forced to adopt a lower bound for the speed c:

$$c^2 \geq 4Dk \tag{2.17}$$

So $c_0 = 2\sqrt{Dk}$ is the *minimal* wave speed (and $e^{-\sqrt{\frac{k}{D}}\theta}$ is the corresponding travelling plane wave solution) and we have found a characterization of the asymptotic speed c_0 which allows for its determination without demanding a prohibitive effort

The following argument due to J.A.J. Metz makes the result intuitively understandable. By manipulating the initial condition suitably we can produce travelling waves in much the same way as one can create the illusion of steady movement in an array of electric lights by turning them on and off appropriately. Only one thing can spoil this game: if we try to make the speed too low the inherent 'infection' mechanism of our excitable medium takes over. Therefore this inherent infection speed is exactly the lowest possible wave speed!

2.2. Host-pathogen systems

Let $S(t,x)$ denote the density of unaffected host plants. For the domain of x (the habitat or field) we simply take $\mathbb{R}^2$. Let $A(\tau,x,y)$ describe the infectivity at x caused by the pathogen on a plant at y which was infected τ time units ago, then, by the law of mass action,

$$\frac{\partial S}{\partial t}(t,x) = S(t,x) \int_0^\infty \int_{\mathbb{R}^2} \frac{\partial S}{\partial t}(t-\tau, y) A(\tau,x,y) dy d\tau. \tag{2.18}$$

If, in the infinite past, S was S_0 (a given function) one obtains upon integrating (2.18) from $-\infty$ to t:

$$u(t,x) = \int_0^\infty \int_{\mathbb{R}^2} g(u(t-\tau, y)) S_0(y) A(\tau,x,y) dy d\tau \tag{2.19}$$

where

$$u(t,x) := -\ln \frac{S(t,x)}{S_o(x)} \tag{2.20}$$

and

$$g(u) = 1 - e^{-u} \tag{2.21}$$

Similarly the equation

$$u(t,x) = \int_0^t \int_{\mathbb{R}^2} g(u(t-\tau, y)) S_0(y) A(\tau,x,y) dy d\tau + f(t,x) \tag{2.22}$$

corresponds to an initial value problem in which at $t=0$ S is given by S_0 and the (given) function f describes the infectivity due to the pathogen already present at $t=0$.

Note that in this model the hosts don't move but the pathogen does by *non-local* interaction (for instance realized by spore dissemination), that an incubation period (time delay between infection and spore production) is incorporated and that the diminution of unaffected hosts makes the problem non-linear. These features create as many striking differences with the diffusion equation of the foregoing section, but nevertheless the description of Section 1.2 reduces both to the same denominator. So let's see whether similar conclusions can be obtained.

We first make two simplifying assumptions:

$$S_0(x) = S_0, \quad \text{a constant,} \tag{2.23}$$

$$A(\tau,x,y) = H(\tau)V(|x-y|). \tag{2.24}$$

The first means that initially the density of unaffected hosts is everywhere the same, the second that the medium for the interaction is homogeneous and isotropic (only the distance between x and y matters; so no prevailing wind) and that the dispersal of infectious agents is so fast relative to the time scale of the incubation and infectivity period that the processes of creation and transport of infectious agents are effectively uncoupled.

If $u(t,x) = w(x \cdot \zeta - ct)$ is to be a solution of (2.19), under the assumption (2.23)-(2.24), the function w has to be a solution of the nonlinear convolution equation on the line

$$w(\theta) = S_0 \int_{-\infty}^{\infty} g(w(\eta)) \tilde{V}_c(\theta - \eta) d\eta, \quad -\infty < \theta < +\infty, \tag{2.25}$$

where

$$\tilde{V}_c(\eta) := \int_0^{\infty} H(\tau) \tilde{V}(\eta - c\tau) d\tau \tag{2.26}$$

with $\tilde{V}$ the so-called marginal infectivity kernel defined by

$$\tilde{V}(\eta) := \int_{-\infty}^{\infty} V(\sqrt{\eta^2 + \sigma^2}) d\sigma. \tag{2.27}$$

In the analysis of (2.25) an important role is played by the characteristic equation

$$L_c(\lambda) = 1 \tag{2.28}$$

where

$$L_c(\lambda) := S_0 \int_{-\infty}^{\infty} \tilde{V}_c(\eta) e^{-\lambda \eta} d\eta = S_0 \int_0^{\infty} e^{-\lambda c \tau} H(\tau) d\tau \int_{\mathbf{R}^2} V(|x|) e^{-\lambda x_1} dx. \tag{2.29}$$

This characteristic equation is obtained by linearizing (2.25) around the constant solution $w \equiv 0$ followed by substitution of an exponential function. Let us assume that both H and V are nonnegative and integrable and that V decreases faster than exponentially for $|x| \to \infty$. Then some straightforward arguments show that the definition

$$c_0 := \inf\{c > 0 | L_c(\lambda) = 1 \text{ for some } \lambda > 0\} \tag{2.30}$$

makes sense (and that $0 < c_0 < \infty$), provided

$$L_c(0) = S_0 \int_0^{\infty} H(\tau) d\tau \int_{\mathbf{R}^2} V(|x|) dx > 1. \tag{2.31}$$

The condition (2.31) is the famous threshold condition of mathematical epidemiology which has the following interpretation: the number of secundary infections produced by a single newly infected individual placed in a hypothetical population (of density S_0) consisting permanently of susceptibles only should exceed one. Clearly any epidemic will peter out immediately if this condition is not satisfied! From now on we assume that (2.31) holds.

THEOREM 1. *For any $c \geqslant c_0$ there exists a nonincreasing solution w of (2.25) with $w(-\infty)=p$ and $w(+\infty)=0$ where p is the unique positive root of the scalar equation*

$$p = \gamma S_0 g(p) \quad \text{where} \quad \gamma := \int_0^\infty H(\tau) d\tau \int_{\mathbb{R}^2} V(|x|) dx.$$

For $c > c_0$, the basic idea of the proof in [20,70] is to use the information obtained from $L_c(\lambda)$ and the properties of g in the construction of two functions ϕ and ψ such that $\phi \leqslant \psi$, $T\phi \geqslant \phi$, $T\psi \leqslant \psi$, where T denotes the (monotone!) integral operator that is associated with the right-hand side of (2.25). For $c = c_0$ one can either follow the same procedure, but the construction is a little bit more complicated, see [70], or one can resort to a limiting argument which shows that the set of speeds is closed, see [13].

The characterization of the set of speeds is completed by the following complementary result.

THEOREM 2. *For $0 \leqslant c < c_0$ there are no nonconstant solutions of equation (2.25) with $0 \leqslant w(\theta) \leqslant p$.*

One can prove Theorem 2 in at least two different ways. In one approach one has to construct a compactly supported function ψ such that, for δ positive and sufficiently small, $T(\delta\psi) \geqslant \delta\psi$ and $\lim\inf_{n\to\infty} T^{(n)}(\delta\psi) \geqslant p$. Subsequently one shows that for an arbitrary nontrivial solution w of (2.25) there exists a positive δ such that $w \geqslant \delta\psi$ and the result $w \geqslant p$ follows from the monotonicity of T; see [70].

In the second approach one uses Tauberian theorems (notably Pitt's form of Wiener's Tauberian Theorem) to deduce that an arbitrary solution of (2.25) with $0 \leqslant w(\theta) \leqslant p$ has to decrease exponentially to zero for $\theta \to +\infty$. Furthermore, by manipulating a bit with Laplace transforms, one can show that the exponent has to be a real root of the characteristic equation (2.28) and consequently the nonexistence of such roots implies the nonexistence of solutions of (2.25) between 0 and p; see [24] for the details.

The advantage of the second approach is that the same method is suitable for obtaining results about uniqueness modulo translation:

THEOREM 3. *For fixed $c \geqslant c_0$ equation (2.25) admits modulo translation one and only one nonconstant solution between 0 and p.*

The case $c>c_0$ is dealt with in [24] but LUI [48] has extended the proof to the case $c=c_0$; BARBOUR [5] has given a different uniqueness proof based on probabilistic arguments.

In conclusion of this section we state two results which together define the sense in which c_0 is the asymptotic speed of propagation of disturbances.

THEOREM 4. *Let f be a nonnegative bounded continuous function from $\mathbb{R}_+ \times \mathbb{R}^2$ into $\mathbb{R}$ such that the projection of the support of f on $\mathbb{R}^2$ is compact. Then*

$$\lim_{t\to\infty}(\sup\{u(t,x)|\ |x|\geqslant ct\})=0$$

for any $c>c_0$, where u is the solution of equation (2.22).

THEOREM 5. *Let f be a nonnegative continuous function then*

$$\liminf_{t\to\infty}(\min\{u(t,x)|\ |x|\leqslant ct\})\geqslant p$$

for any $c\in(0,c_0)$, provided f is not identically zero.

The proofs are based on a comparison principle and the construction of suitable upper- and lower-solutions [21,67,68]. An understanding of the way in which Volterra convolution equations generate dynamical systems [23] is very helpful.

So, with the part of ∞ assigned to p, a dynamical picture emerges that is identical to the one of the linear diffusion equation.

2.3. Into the field

As presented in Section 2.2 the results have hardly any appeal to researchers in plant pathology. The functions H and V are introduced in the abstract and the theorems are completely unreadable. In an attempt to bridge the communication gulf J.A.J. Metz asked F. van den Bosch, at that time a student in theoretical biology at the University of Leiden, to learn both languages and act as an interpreter. This is a far from easy job but several recent preprints witness that the attempt was quite succesful [9,10]. In joint work with J.C. Zadoks of the Laboratory for Phytopathology of the Agricultural University of Wageningen they developed several mechanistic submodels for spore dispersal from which V can be derived, they introduced flexible yet parameter sparse kernels H that fit published data on spore production well, they developed approximation formulae and numerical procedures to calculate c_0 from the defining equations

$$L_c(\lambda)=1,\quad \frac{\partial L_c}{\partial \lambda}(\lambda)=0$$

with a pocket calculator in negligible time, they expressed both the 'input' quantities S_0, H and V and the 'output' quantity c_0 in standard phytopathological terminology and, finally they showed that the model predictions match up to simulation studies [72] and agree reasonably with the speed measured in a field experiment. They built the connection between some parts of the biological and the mathematical world by making biologically palpable what is mathematically so easily introduced ('Let H and V denote ...').

So far their work deals with the expansion of a connected area of infested plants within a field (a focus or hot-spot). But, as HEESTERBEEK [38] has described and classified in detail, one can consider the spread of an infectious disease in a crop at different geographical scales. One can concentrate on focus expansion, on changes in the number and size of foci within one field or on a large number of fields in different phases of disease development. In the first two cases the temporal scale is the growing season but in the last case one may have to pay attention to overwintering. This last case is particularly relevant in view of so-called quarantine-diseases (pests which are accidentally introduced in countries or continents in which they were unknown before). Although from a mathematical point of view the phenomena are almost identical on all these scales, it is a far from trivial modelling problem to make the available results applicable to the various situations and to figure out what additional results are needed. Work on these problems is in progress.

2.4. Some history and other things worth knowing

The subject of a wave-like transition from an unstable state to a stable one seems to be born in 1937 with the publication of two highly influential papers.

In his paper 'The wave of advance of advantageous genes' [30] FISHER discusses the nonlinear diffusion equation on the line

$$\frac{\partial u}{\partial t} = D\frac{\partial^2 u}{\partial x^2} + f(u)$$

with $f(u) = ku(1-u)$ and he finds that travelling waves exist for all $c \geq c_0 = 2\sqrt{Dk}$. A little puzzled by the indeterminacy of velocity he examines the behaviour of a finite aggregate of discrete particles, subject to random scattering and increase in number, and concludes from this study that c_0 has to be the 'true' speed. In a celebrated paper of the same year 1937 KOLMOGOROFF, PETROVSKY and PISCOUNOFF [46] prove that the solution corresponding to the special discontinuous initial condition

$$u(0,x) = \begin{cases} 0 & x < 0 \\ 1 & x \geq 0 \end{cases}$$

converges to the travelling wave w of minimal velocity c_0 in the sense that $u(t, x+m(t)) \to w(x)$, uniformly in x, for $t \to \infty$ and for appropriate choice of $m(t)$, and that $\dot{m}(t) \to c_0$. Already in 1948 KENDALL [43] observes that this result cannot hold for all initial conditions, but that it is likely that for compactly supported initial data the solution develops into two diverging travelling waves of minimal velocity. Since that time several important contributions to the solution of the convergence problem have been made by various authors, culminating in a complete solution by M. BRAMSON [11] which, remarkably, uses the Feyman-Kac integral formula in conjunction with sample path estimates for Brownian motion as the basic technical device. No results about convergence to travelling waves in higher dimensional spatial domains seem to be known.

But the speed-ambiguity which annoyed Fisher was fully resolved in 1975 when ARONSON and WEINBERGER [2,3] introduced the notion of the *asymptotic speed* of propagation of disturbances and showed that, even in higher space dimension, this speed coincides with the minimal velocity of travelling plane waves. The papers by Aronson and Weinberger mark the beginning of an explosive increase in published papers on nonlinear reaction-diffusion equations with biological applications, see for instance FIFE [27,28], OKUBO [55] and DIEKMANN and TEMME [26].

As the title of his paper indicates, Fisher was interested in the speed at which an advantageous mutant gene would spread in a spatially distributed population. In a similar spirit SKELLAM [63] investigated the regional spread of oak trees in the post-glacial period and the dispersal of the muskrat after its escape from 'prison' in Europe, and AMMERMAN and CAVALLI-SFORZA [1] analysed the neolithic transition in Europe (the shift from hunting and gathering to early farming as a new way of life). KENDALL [44] initiated the modelling of the spatial spread of epidemics (his work has been continued by MOLLISON [52]). As a rather sinister example NOBLE [54] has treated the propagation of the Black Death in medieval Europe. A much studied wildlife disease is rabies [4].

The model of Section 2.2 is a space-dependent analogue of the basic model of KERMACK and MCKENDRICK (which was introduced as early as 1927 [45]; also see [50]). It was developed and analysed independently by THIEME [65] and DIEKMANN [20] and later extended to vector-borne and other multi-type diseases by RADCLIFFE and RASS [60]. A remarkable feature of both the epidemic equation and the nonlinear diffusion equation with $f(u)=ku(1-u)$ is that c_0 is determined by the linearization at the unstable state. This is true for a large class of nonlinearities but not for all (in this connection one discriminates between pulled waves, the ones we have met, and pushed waves which are more strongly determined by the nonlinearity; see, for instance, ROTHE [61], HADELER and ROTHE [35]).

If in the genetics model heterozygotes are inferior one has *two* steady states which are 'seperated' from one another by an unstable steady state. In this case there exists usually a unique (modulo translation) wave travelling at an exactly determined velocity. In order to bring about a transition from one stable state to the other perturbations now have to be sufficiently large over a sufficiently large domain (super-threshold, as it is called) but once this is so the transition takes effectively place with the wave velocity, see FIFE and MCLEOD [29] and [27,28].

WEINBERGER [71] has introduced and analysed a discrete time equation which is sufficiently general to cover both discrete and continuous spatial domains and which allows for seasonal influences and spatial anisotropy (prevailing winds!). As a consequence the speed may depend on the direction. Let $c_0(\zeta)$ be the minimal speed of travelling plane waves in the direction ζ then Weinberger shows that the (convex) set

$$S = \{x \in \mathbb{R}^2 | x \cdot \zeta \leq c_0(\zeta) \text{ for all unit vectors } \zeta\}$$

replaces the circles in the results that characterize the asymptotic speed of propagation. Many other results for this class of equations were obtained by LUI in an interesting series of papers [48,49].

Aronson and Weinberger have achieved a major conceptual break-through by introducing the notion of 'asymptotic speed of propagation'. This notion combines practical relevance with mathematical elegance. Analysis of a multitude of models has by now made clear that it provides a robust link between observed spatial expansion of many different substances and the behaviour of solutions of mathematical equations. The characterization as the minimal wave speed makes it computable and hence applicable.

It is not always easy to apply applied mathematics. The spirit of the papers by Fisher and by Aronson and Weinberger is quite different and so is the jargon. The style of the papers by Thieme and Diekmann puts off many potentially interested people. We need chains of communicating people with overlapping knowledge and interests to let the stream of scientific information and inspiration flow freely back and forth between scientists and mathematicians. In Section 2.3 I briefly described such a chain and indicated its highly valuable products.

The early papers (FISHER [30], SKELLAM [63]) are quite explicitly concerned with natural phenomena. Next comes a period in which 'applicability' is still a motivation, but nevertheless mathematical analysis is the principal thing. The right concept is created and strong results are obtained. It requires additional energy to come full circle and let the mathematical results bear upon the original scientific questions. Most likely new questions arise in this 'final' phase and the process repeats indefinitely ('the march of science along a spiral staircase').

3. MATHEMATICAL MODELS OF STRUCTURED POPULATIONS AND PERTURBED DUAL SEMIGROUPS

3.1. The background

The first impulse to a general theory of physiologically structured population models was given in 1967, a year which showed a remarkable outburst of innovative papers [8,31,62]. But, perhaps due to the lack of a cut and dried mathematical framework, the subsequent development was disappointing in view of the very promising start. In the first half of 1983 a colloquium on the Dynamics of Structured Populations was held at the Centre for Mathematics and Computer Science attempting to revive the spirit of the pioneering papers and, at the same time, to start building the required mathematical framework. The colloquium served as a starting point for intense interdisciplinary interaction of the core participants. The fruits of this interaction obtained so far have been documented extensively elsewhere [51]. Here I want to concentrate on one particular mathematical aspect while refering to [51] for a general survey and many concrete examples displaying various amounts of biological complexity and realism.

3.2. Age-dependent population growth

Let the individuals of a population be characterized by their age a. Let $n(t,a)$ denote the age distribution at time t, i.e.

$$\int_{a_1}^{a_2} n(t,\alpha)d\alpha = \text{number of individuals with age between } a_1 \text{ and } a_2 \text{ at time } t.$$

The individuals age, may give birth or die. The first process is described by the differential equation $\frac{da}{dt}=1$, the second by the age-specific per capita birth rate $\beta(a)$ and the third by the age-specific per capita death rate $\mu(a)$. Since

$$n(t+h,a+h) = n(t,a) - h\mu(a)n(t,a) + O(h^2)$$

we derive for n the balance law

$$\frac{\partial n}{\partial t} = -\frac{\partial n}{\partial a} - \mu n \tag{3.1}$$

which we supplement with the boundary condition

$$n(t,0) = \int_0^\infty \beta(\alpha)n(t,\alpha)d\alpha \tag{3.2}$$

to express that the influx at the boundary $a=0$ equals the total birth rate. Finally we assume that at $t=0$ the age distribution equals a given function ϕ:

$$n(0,a) = \phi(a). \tag{3.3}$$

In order to minimize inessential (for the present purpose) technical and notational detail we take μ to be identically zero throughout this paper. To get a feel for the problem we begin by taking $\beta(a) \equiv 0$ as well. In the absence of births and deaths the solution of (3.1) - (3.3) is evidently

$$n(t,a,\phi) = \begin{cases} \phi(a-t) &, a \geq t \\ 0 &, a < t \end{cases} \tag{3.4}$$

as follows also directly from the interpretation.

A reasonable choice of population state space is $L_1(\mathbb{R}_+)$. Putting

$$T_0(t)\phi = n(t,\cdot,\phi) \tag{3.5}$$

we obtain a strongly continuous semigroup of bounded linear operators on $L_1(\mathbb{R}_+)$ with infinitesimal generator

$$\begin{cases} A_0\phi = -\phi' \\ D(A_0) = \{\phi | \phi(a) = \int_0^a \phi'(\alpha)d\alpha \text{ with } \phi' \in L_1(\mathbb{R}_+)\} \end{cases} \tag{3.6}$$

(recalling that one out of several equivalent definitions of an absolutely continuous function is 'a function which is, locally, the integral of an L_1-function', we can also write $D(A_0) = \{\phi | \phi \text{ is absolutely continuous, } \phi(0)=0 \text{ and } \phi'$ is integrable over $\mathbb{R}_+$}; in the following we abbreviate 'absolutely continuous' to AC).

The standard solution procedure in case of non-zero birth rate is the following. First consider the birth rate

$$b(t) = \int_0^\infty \beta(\alpha) n(t,\alpha) d\alpha \tag{3.7}$$

as known. Then

$$n(t,a,\phi) = \begin{cases} \phi(a-t) & , a \geq t \\ b(t-a) & , a < t \end{cases} \tag{3.8}$$

where, although we haven't expressed this in our notation, b depends on ϕ. Substituting (3.8) into (3.2) we obtain the linear renewal (i.e. Volterra convolution) equation

$$b(t) = \int_0^t \beta(\alpha) b(t-\alpha) d\alpha + f(t) \tag{3.9}$$

with

$$f(t) = \int_t^\infty \beta(\alpha) \phi(\alpha - t) d\alpha = \int_0^\infty \beta(\alpha + t) \phi(\alpha) d\alpha. \tag{3.10}$$

Assume $\beta \in L_\infty(\mathbb{R}_+)$. Standard contraction mapping arguments imply that (3.9) has a unique solution represented by

$$b = \sum_{n=0}^\infty \beta^{n*} \star f \tag{3.11}$$

where the star denotes the convolution product, $\beta^{0*} \star f := f$, $\beta^{1*} := \beta$, $\beta^{n*} := \beta^{(n-1)*} \star \beta$, $n \geq 2$. Substituting (3.11) into (3.8) we finally arrive at a series expansion for the age distribution n, which has the following interpretation. Let's call those individuals which were present at time $t=0$ the zero'th generation. Then f describes the offspring of the zero'th generation and the corresponding term in the expansion of n is, for this reason, called the first generation. Similarly the n-th term describes the n-th biological generation and the expansion is called the generation expansion.

The semigroup

$$T(t)\phi = n(t,\cdot,\phi) \tag{3.12}$$

is generated by

$$\begin{cases} A\phi = -\phi' \\ D(A) = \{\phi | \phi(a) = \int_0^\infty \beta(\alpha)\phi(\alpha) d\alpha + \int_0^a \phi'(\alpha) d\alpha \text{ with } \phi' \in L_1(\mathbb{R}_+)\} \end{cases} \tag{3.13}$$

(or, equivalently, $D(A) = \{\phi | \phi \text{ is AC}, \phi(0) = \int_0^\infty \beta(\alpha)\phi(\alpha)d\alpha \text{ and } \phi' \text{ is integrable over } \mathbb{R}_+\}$).

A striking point is that *all* information about the birth rate enters in the domain of A and that the action of A is independent of β. This is highly unpleasant for several reasons:

(i) within the present functional analytic framework there is no analogue of the renewal equation (3.9) which we can solve iteratively; a puzzling and somewhat irritating phenomenon.

(ii) if we deal with nonlinear birthrates (describing density dependence) we don't have at our disposal a variation-of-constants formula. The lack of this important tool forms an obstacle for the development of the local stability and bifurcation theory and as a consequence ad hoc approaches dominate the field [69,59,19].

So, once again, is an abstract approach beneficial? It looks as though we made life more complicated, instead of simpler, by introducing a semigroup.

A little reflection reveals that the difficulty is due to the fact that all newborns have (by definition) one and the same age $a = 0$. The range of the birth operator is spanned by the (Dirac) measure concentrated at $a = 0$ which is not an element of $L_1(\mathbb{R}_+)$. So the 'perturbation' of the generator maps out of the state space into some bigger space but, as we have seen, solving the differential equation we come back into the smaller space. An analogous phenomenon occurs with delay equations [36,22].

Should we enlarge the state space and let our age distributions live in the space of regular Borel measures? This is a natural and sensible action (in fact one can argue right from the start that this is the appropriate state space) but we have to pay a technical price: the semigroup is no longer strongly continuous (indeed, translation of a concentrated measure is not continuous).

In Section 3.4 it is shown that we need not choose the least of two evils but that, instead, we can make great play with the good things of two spaces neither of which is ideal by itself. It appears that duality provides us with a systematic procedure to create the appropriate 'bigger' space and that a general theory can be built which encompasses both age-dependent population models and delay equations. The key Section 3.4 is essentially a summary of the preprint [16] by CLÉMENT, DIEKMANN, GYLLENBERG, HEIJMANS and THIEME.

3.3. Dual semigroups

Let $\{T(t)\}$ be a strongly continuous semigroup of bounded linear operators on a Banach space X generated by A. The adjoint operators $T^*(t)$ form a semigroup on the dual space X^*. $\{T^*(t)\}$ is weak $*$ continuous but need not be strongly continuous if we equip X^* with the norm topology (unless X is reflexive). A^*, the adjoint of A, is the weak $*$ generator of $\{T^*(t)\}$. Note that A^* need not be densely defined.

In their classic treatise [42] HILLE and PHILLIPS showed that the dialogue of a space and a semigroup demands a duality theory which is made to measure.

We need a special star, called sun and represented by the symbol $\odot$. Let $X^\odot$ denote the maximal invariant subspace on which $\{T^*(t)\}$ is strongly continuous. Then

$$X^\odot = \{\phi^* \in X^* \mid \lim_{t \downarrow 0} \|T^*(t)\phi^* - \phi^*\| = 0\}, \tag{3.14}$$

$X^\odot$ is norm-closed and $\overline{D(A^*)} = X^\odot$. Let $\{T^\odot(t)\}$ denote the strongly continuous semigroup on $X^\odot$ which is obtained by restriction of $\{T^*(t)\}$ and let $A^\odot$ denote its generator. Then $A^\odot$ is the part of A^* in $X^\odot$, i.e. the largest restriction of A^* with both domain and range in $X^\odot$.

On $X^{\odot*}$, the dual space of $X^\odot$, we obtain a weak $*$ continuous semigroup $\{T^{\odot*}(t)\}$ with weak $*$ generator $A^{\odot*}$. Let

$$X^{\odot\odot} = \{\phi^{\odot*} \in X^{\odot*} \mid \lim_{t \downarrow 0} \|T^{\odot*}(t)\phi^{\odot*} - \phi^{\odot*}\| = 0\}. \tag{3.15}$$

It follows rather easily that X can be embedded into $X^{\odot*}$ and henceforth we identify X and its embedding. Then X becomes a subspace of $X^{\odot\odot}$.

DEFINITION. *X is called $\odot$-reflexive with respect to A iff $X = X^{\odot\odot}$.*

It is known that X is $\odot$-reflexive with respect to A iff $(\lambda I - A)^{-1}$ is $X^\odot$-weakly compact. Moreover, X is $\odot$-reflexive with respect to A iff $X^\odot$ is $\odot$-reflexive with respect to $A^\odot$.

3.4. Perturbation theory for dual semigroups
Let $\{T_0(t)\}$ be a strongly continuous semigroup on X generated by A_0 and assume that X is $\odot$-reflexive with respect to A_0. Let $B: X \to X^{\odot*}$ be a bounded linear operator. The variation-of-constants equation

$$T(t)\phi = T_0(t)\phi + \int_0^t T_0^{\odot*}(t-\tau)BT(\tau)\phi \, d\tau \tag{3.16}$$

can be shown to make sense and to admit a unique solution $\{T(t)\}$ (which can be represented by a 'generation' series). Here the integral is a weak $*$ integral which in principle takes values in $X^{\odot*}$ but in fact takes values in the closed subspace $X^{\odot\odot} = X$. By duality and restriction we obtain semigroups $\{T^*(t)\}$, $\{T^\odot(t)\}$ and $\{T^{\odot*}(t)\}$ on $X^*, X^\odot$ and $X^{\odot*}$ respectively, since it can be shown that the spaces of strong continuity do *not* depend on B. Similarly the domains of the weak $*$ generators on the 'big' spaces are independent of B. The following theorem summarizes part of the results.

THEOREM. *The operator $A\phi = A_0^{\odot*}\phi + B\phi$ with $D(A) = \{\phi \in D(A_0^{\odot*}) \mid A_0^{\odot*}\phi + B\phi \in X\}$ is the generator of a strongly continuous semigroup $\{T(t)\}$ on X and the variation-of-constants formula (3.16) holds.*

The symmetry of the framework is apparent from the diagram

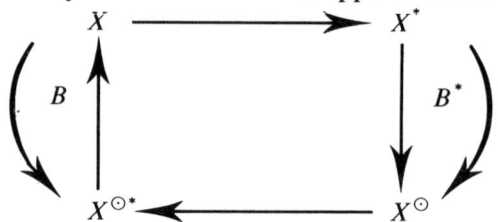

When X is not $\odot$-reflexive with respect to A_0 this symmetry is disturbed. Nevertheless similar results hold. A canonical embedding of $X^{\odot\odot}$ into X^{**} seems to play a leading part, but it is not yet precisely clear how the most elegant and efficient argumentation proceeds, so we refrain from further discussion here.

3.5. Age-dependent population dynamics revisited

If we consider age-distributions over the non-compact domain $\mathbb{R}_+$ we don't get $\odot$-reflexivity. However, if β has compact support (or, in other words, if very old individuals don't produce offspring) we may limit our bookkeeping of individuals without losing relevant information. For the purpose of the present exposition, we therefore replace $L_1(\mathbb{R}_+)$ by $L_1(0,a_{max})$ for some constant a_{max}. So $X = L_1(0,a_{max})$.

Let $A_0\phi = -\phi'$ with $D(A_0) = \{\phi | \phi \text{ is AC and } \phi(0)=0\}$ be, as before, the generator of the semigroup

$$(T_0(t)\phi)(a) = \begin{cases} \phi(a-t) &, a \geq t \\ 0 &, a < t. \end{cases}$$

On the dual space $X^* = L_\infty(0,a_{max})$ we have the semigroup

$$(T_0^*(t)\psi)(a) = \begin{cases} \psi(a+t) &, a+t \leq a_{max} \\ 0 &, a+t > a_{max} \end{cases}$$

with weak * generator

$A_0^*\psi = \psi'$
$D(A_0^*) = \{\psi | \psi \text{ has a Lipschitz continuous representative which is zero at } a = a_{max}\}$.

Clearly $X^\odot = C_0(0,a_{max}) = \{\psi | \psi \text{ has a continuous representative which is zero at } a = a_{max}\}$ and $X^{\odot *} = M[0,a_{max}]$, the space of all complex regular Borel measures on $[0,a_{max})$. It is well-known [14] that the subspace of $M[0,a_{max})$ on which translation is continuous, i.e. $X^{\odot\odot}$, is exactly the closed subspace of all absolutely continuous measures. The mapping which associates with any ϕ in X the measure μ in $X^{\odot *}$ defined by $\mu(\omega) = \int_\omega \phi(\alpha)d\alpha$ describes the canonical identification of X and $X^{\odot\odot}$.

Let $B: X \to X^{\odot *}$ be defined by

$$B\phi = \int_0^{a_{max}} \beta(\alpha)\phi(\alpha)d\alpha \cdot \delta = \langle \beta, \phi \rangle \delta$$

Dynamics in bio-mathematical perspective 45

where δ is the (Dirac) measure concentrated at $a=0$. Then the results of Section 3.4 apply and we conclude that the part of $A_0^{\odot *}+B$ in X generates a semigroup $T(t)$ which satisfies the variation-of-constants equation

$$T(t)\phi = T_0(t)\phi + \int_0^t T_0^{\odot *}(t-\tau)BT(\tau)\phi d\tau. \tag{3.17}$$

Since B has one-dimensional range we can go a little further. Define $b(t) = \langle \beta, T(t)\phi \rangle$ then, applying the functional β to (3.17), we find after a little calculation that b has to satisfy the scalar equation

$$b(t) = f(t) + \int_0^t \beta(t-\tau)b(\tau)d\tau \tag{3.18}$$

where $f(t) := \langle \beta, T_0(t)\phi \rangle = \int_0^{a_{max}} \beta(\alpha)\phi(\alpha-t)d\alpha$. Thus we recover the renewal equation (3.9).

If, conversely, b is a solution of (3.18) with f of the form $f(t) = \langle \beta, T_0(t)\phi \rangle$ for some $\phi \in X$ then $T(t)\phi$ is obtained by a simple substitution into the now explicit expression (3.17):

$$T(t)\phi = T_0(t)\phi + \int_0^t T_0^{\odot *}(t-\tau)\delta b(\tau)d\tau \tag{3.19}$$

Thus we obtain a reformulation of (3.8).

We conclude that the abstract framework of Sections 3.3 and 3.4 is rich enough for the (re)formulation of the (quasi-) explicit formulas of the direct approach via the renewal equation.

Now we can also ease those attentive readers who worried about the fact that the epidemic model of Section 2.2 was formulated as an integral equation (notably with respect to the time variable) and not as a (abstract) differential equation. When we think of 'age' as 'time elapsed since infection' and adopt a nonlinear 'birth = infection' condition one can make the connection between the nonlinear renewal equation via the variation-of-constants formula (3.17) exactly as in the present linear case.

3.6. Physiologically structured population models: a challenge for the future
The biological motivation for studying physiologically structured population models is described at length in the lecture notes [51] and the survey paper [64]. The mathematical form taken by these models is:

$$\frac{\partial n}{\partial t} + \text{divergence (velocity } n) = \text{sources} - \text{sinks}, \quad x \in \Omega,$$

$$\nu \cdot \text{velocity}|_{\partial \Omega_+} = \text{source}$$

where the individual 'velocity' $\frac{dx}{dt}$ and the sources and sinks are specified according to the specific situation at hand. Here ν denotes the inward normal to $\partial\Omega$, the boundary of the individual state space Ω, and $\partial\Omega_+$ is the part of $\partial\Omega$

at which $v \cdot$ velocity >0, i.e. characteristics enter Ω. The solution concept is based on integration along characteristics.

In a recent survey on *Infinite Dimensional Dynamics* [37] J.K. HALE writes:

> 'For the successful development and application of dynamical systems in infinite dimensions, we need intensive interaction between two special groups of researchers. The first group consists of mathematicians who are well trained in dynamical systems and know both the analytic and the geometric theory of differential equations in finite dimensions. They should also know well the classical and modern theory of partial differential and functional differential equations and have a strong background in applications-especially physics and engineering. The other group of researchers should be primarily concerned with applications, but should be well trained in ordinary and partial differential equations. It does not take much reflection to see that there are very few people with these qualifications. More resources need to be allocated for training young people to carry out this program'.

Then Hale goes on to describe functional differential equations and parabolic systems as special cases in which the type of interaction he has in mind has led to considerable success (and to make some remarks about hyperbolic systems and chaotic dynamics). It seems quite conceivable that the equations of physiologically structured population dynamics will be at home in a similar survey written many years from now. But whether this will happen or not, only time will tell.

REFERENCES

1. A.J. AMMERMAN, L.L. CAVALLI-SFORZA (1984). *The Neolithic Transition and the Genetics of Populations in Europe*, Princeton Univ. Press.
2. D.G. ARONSON, H.F. WEINBERGER (1975). Nonlinear diffusion in population genetics, combustion and nerve pulse propagation. J.A. GOLDSTEIN (ed.). *Partial Differential Equations and Related Topics*, Springer Lect. Notes in Math. 446, 5-49.
3. D.G. ARONSON, H.F. WEINBERGER (1978). Multidimensional nonlinear diffusion arising in population genetics. *Adv. in Math.* 30, 33-76.
4. P.J. BACON (ed.). (1985). *Population Dynamics of Rabies in Wildlife*, Academic Press.
5. A.D. BARBOUR (1977). The uniqueness of Atkinson and Reuter's epidemic waves. *Math. Proc. Camb. Phil. Soc.* 82, 127-130.
6. V. BARBU (1976). *Nonlinear Semigroups and Differential Equations in Banach Spaces*, Noordhoff, Leiden.
7. G.I. BARENBLATT (1979). *Similarity, Self-Similarity and Intermediate Asymptotics*, Plenum.

8. G.I. BELL, E.C. ANDERSON (1967). Cell growth and division. I. A mathematical model with applications to cell volume distributions in mammalion suspension cultures. *Biophys. J. 7,* 329-351.
9. F. VAN DEN BOSCH, J.A.J. METZ, J.C. ZADOKS. *The Asymptotic Speed of Travelling Epidemic Waves,* preprint
10. F. VAN DEN BOSCH, J.C. ZADOKS, J.A.J. METZ. *Focus Formation in Plant Diseases.* I. The constant rate of focus expansion. II. Realistic parameter-sparse models. Preprints.
11. M. BRAMSON (1983). Convergence of solutions of the Kolmogorov equation to travelling waves. *Memoir of the AMS 285.*
12. H. BREZIS (1977). *Opérateurs Maximaux Monotones et Semi-Groupes de Contractions dans les Espaces de Hilbert,* North Holland, Amsterdam.
13. K.J. BROWN, J. CARR (1977). Deterministic epidemic waves of critical velocity. *Math. Proc. Camb. Phil. Soc. 81,* 431-436.
14. P.L. BUTZER, H. BERENS (1967). *Semi-groups of Operators and Approximation,* Springer, Berlin.
15. J. VAN CASTEREN (1985). *Generators of Strongly Continuous Semigroups,* Pitman, Boston.
16. PH. CLÉMENT, O. DIEKMANN, M. GYLLENBERG, H.J.A.M. HEIJMANS, H.R. THIEME. *Perturbation Theory for Dual Semigroups. I. The Sun-Reflexive Case,* preprint.
17. M.G. CRANDALL (1986). Nonlinear semigroups and evolution governed by accretive operators. *Proc. Symp. Pure Math. AMS 45* Part 1, 305-337.
18. E.B. DAVIES (1980). *One-Parameter Semigroups,* Academic Press, London.
19. W. DESCH, W. SCHAPPACHER (1985). Spectral properties of finite-dimensional perturbed linear semigroups. *J. Diff. Equ. 59,* 80-102.
20. O. DIEKMANN (1978). Thresholds and travelling waves for the geographical spread of infection. *J. Math. Biol. 6,* 109-130.
21. O. DIEKMANN (1979). Run for your life. A note on the asymptotic speed of propagation of an epidemic. *J. Diff. Equ. 33,* 58-73.
22. O. DIEKMANN. *Perturbed Dual Semigroups and Delay Equations,* preprint.
23. O. DIEKMANN, S.A. VAN GILS (1984). Invariant manifolds for Volterra integral equations of convolution type. *J. Diff. Equ. 54,* 139-180.
24. O. DIEKMANN, H.G. KAPER (1978). On the bounded solutions of a nonlinear convolution equation. *Nonl. Anal. Th. Math. Appl. 2,* 721-737.
25. O . DIEKMANN, J.A.J. METZ, S.A.L.M. KOOIJMAN, H.J.A.M. HEIJMANS (1984). Continuum population dynamics with an application to *Daphnia magna. Nieuw Archief voor Wiskunde 4,* 82-109.
26. O. DIEKMANN, N.M. TEMME (1976). *Nonlinear Diffusion Problems, MC Syllabus 28,* Math. Centrum, Amsterdam.
27. P.C. FIFE (1978). Asymptotic states for equations of reaction and diffusion. *Bull. AMS 84,* 693-726.
28. P.C. FIFE (1979). *Mathematical Aspects of Reacting Diffusing Systems,* Springer Lect. Notes in Biomath. 28.
29. P.C. FIFE, J.B. MCLEOD (1977). The approach of solutions of nonlinear diffusion equations to travelling front solutions. *Arch. Rat. Mech. Anal.*

65, 335-361.
30. R.A. FISHER (1937). The wave of advance of advantageous genes. *Ann. of Eugenics 7*, 355-369.
31. A.G. FREDRICKSON, D. RAMKRISHNA, H.M. TSUCHIYA (1967). Statistics and dynamics of procaryotic cell populations. *Math. Biosc. 1*, 327-374.
32. A. FRIEDMAN (1969). *Partial Differential Equations*, Holt-Rinehart & Winston.
33. J.A. GOLDSTEIN (1985). *Semigroups of Operators and Applications*, Oxford University Press.
34. M.E. GURTIN, R.C. MACCAMY (1974). Nonlinear age-dependent population dynamics. *Arch. Rat. Mech. Anal. 54*, 281-300.
35. K.P. HALDELER, F. ROTHE (1975). Travelling fronts in nonlinear diffusion equations. *J. Math. Biol. 2*, 251-263.
36. J.K. HALE (1977). *Theory of Functional Differential Equations*, Springer.
37. J.K. HALE (1985). *Infinite Dimensional Dynamics*, Report, Brown Univ. Providence. R.I.
38. H. HEESTERBEEK (1985). *Over Modellering van Continentale Epidemieën*, Laboratory of Phytopathology, Agricultural University Wageningen.
39. D. HENRY (1981). *Geometric Theory of Semilinear Parabolic Equations, Springer Lect. Notes. in Math. 840.*
40. H.J.A.M. HEIJMANS (1985). *Dynamics of Structured Populations*, Thesis, Univ. of Amsterdam.
41. H.J.A.M. HEIJMANS (1986). Structured populations, linear semigroups and positivity. *Math. Z. 191*, 599-617.
42. E. HILLE, R.S. PHILLIPS (1957). *Functional Analysis and Semi-Groups*, Amer. Math. Soc., Providence R.I.
43. D.G. KENDALL (1948). A form of wave propagation associated with the equation of heat conduction. *Proc. Camb. Phil. Soc 44*, 591-593.
44. D.G. KENDALL (1965). Mathematical models of the spread of infection. *Mathematics and Computer Science in Biology and Medicine*, Medical Research Councel, London 213-224.
45. W.O. KERMACK, A.G. MCKENDRICK (1927). A contribution to the mathematical theory of epidemics. *Proc. Roy. Soc. A 115*, 700-721.
46. A. KOLMOGOROFF, I. PETROVSKY, N. PISCOUNOFF (1937). Étude de l'equation de la diffusion avec croissance de la quantité de matière et son application a un problème biologique. *Bull. Univ. Etat Moscou Ser. Int. A. Math. Méc. 1 # 6*, 1-25 *(Bjal. Moskovskovo Gos. Univ. 17*, 1-72).
47. C.C.LIN, L.A. SEGEL (1974). *Mathematics Applied to Deterministic Problems in the Natural Sciences*, Macmillan, New York.
48. R.LUI. A nonlinear integral operator arising from a model in population genetics. *SIAM J. Math. Anal. I* Monotone initial data. *13* (1982), 913-937. II Initial data with compact support. *13* (1982), 938-953. III Heterozygote inferior case. *16* (1985), 1180-1206. IV Clines. *17* (1986), 152-168.
49. R. LUI (1983). Existence and stability of travelling wave solutions of a

nonlinear integral operator. *J. Math. Biol. 16,* 199-220.
50. J.A.J. METZ (1978). The epidemic in a closed population with all susceptibles equally vulnerable; some results for large susceptible populations and small initial infections. *Acta Biotheor. 27,* 75-123.
51. J.A.J. METZ, O. DIEKMANN (eds.). *Dynamics of Physiologically Structured Populations, Springer Lecture Notes in Biomathematics,* to appear in 1986.
52. D. MOLLISON (1977). Spatial contact models for ecological and epidemic spread. *J. Roy. Stat. Soc. B 39,* 283-326.
53. R. NAGEL (ed.) (1986). *One-Parameter Semigroups of Positive Operators, Springer Lect. Notes in Math. 1184.*
54. J.V. NOBLE (1974). Geographic and temporal development of plagues. *Nature 250,* 726-729.
55. A. OKUBO (1980). *Diffusion and Ecological Problems: Mathematical Models, Biomathematics Vol. 10,* Springer.
56. A. PAZY (1983). *Semigroups of Linear Operators and Applications to Partial Differential Equations,* Springer, New York.
57. J. PRÜSS (1981). Equilibrium solutions of age-specific population dynamics of several species. *J. Math. Biol. 11,* 65-84.
58. J. PRÜSS (1983). On the qualitative behaviour of populations with age-specific interactions. *Comp. & Maths. with Appls. 9,* 327-339.
59. J. PRÜSS (1983). Stability analysis for equilibria in age-specific population dynamics. *Nonl. Anal. Th. Math. Appl. 7,* 1291-1313.
60. J. RADCLIFF, L. RASS (1984). The spatial spread and final size of the deterministic non-reducible *n*-type epidemic. *J. Math. Biol. 19,* 309-327.
61. F. ROTHE (1981). Convergence to pushed fronts. *Rocky Mountain J. Math. 11,* 617-633.
62. J.W. SINKO, W. STREIFER (1967). A new model for age-size structure of a population. *Ecology 48,* 910-918.
63. J.G. SKELLAM (1951). Random dispersal in theoretical populations. *Biometrika 38,* 196-218.
64. W.STREIFER (1974). Realistic models in population ecology. A. MAC FADYEN (ed.). *Advances in Ecological Research 8,* 199-266.
65. H.R. THIEME (1977). A model for the spatial spread of an epidemic. *J. Math. Biol. 4,* 337-351.
66. H.R. THIEME (1977). The asymptotic behaviour of solutions of nonlinear integral equations. *Math. Z. 157,* 141-154.
67. H.R. THIEME (1979). Asymptotic estimates of the solutions of nonlinear integral equations and asymptotic speeds for the spread of populations. *J. Reine Angew. Math. 306,* 94-121.
68. H.R. THIEME (1979). Density-dependent regulation of spatially distributed populations and their asymptotic speed of spread. *J. Math. Biol. 8,* 173-187.
69. G.F. WEBB (1985). *Theory of Nonlinear Age-Dependent Population Dynamics,* Marcel Dekker.
70. H.F. WEINBERGER (1978). Asymptotic behaviour of a model in population genetics. J.M. CHADAM (ed.). *Nonlinear Partial Differential Equations and*

Applications, Springer Lect. Notes in Math. *648,* 47-98.
71. H.F. WEINBERGER (1982). Long-time behaviour of a class of biological models. *SIAM J. Math. Anal. 13,* 353-396.
72. J.C. ZADOKS, P. KAMPMEIJER (1977). *Epimul, a Simulator of Foci and Epidemics in Mixtures of Resistant and Susceptible Plants, Mosaics and Multilines,* simulation monograph, PUDOC, Wageningen.

De Erfvijand Wiskundig Bestreden

(The Arch-enemy Attacked Mathematically)

L. de Haan
Erasmus University Rotterdam
P.O. Box 1738, 3000 DR Rotterdam, The Netherlands

INTRODUCTION

Approximately 40% of the Netherlands is below sea-level and has to be protected against the sea by dikes. No specific statistical study was done to fix a safer level for the sea-dikes before 1953. On February 1, 1953 during a severe windstorm combined with high tide in several parts of the Netherlands (mainly Holland and Zeeland) the sea-dikes broke, part of the country was flooded and nearly two thousand people were killed. The breaking of the dikes was caused by the unpredicted high level of the North Sea at that particular time and place: the water went over the dike, the unconsolidated backside was gradually washed away and finally the dike collapsed.

Since it was apparent that the sea-dikes were too low, the government appointed a committee (the so-called Delta-committee) to recommend on an appropriate level for the dikes (called Delta-level since). A statistical group from the Mathematical Centre in Amsterdam headed by professor D. VAN DANTZIG went to work and came up with (approximately) the following solution of the problem. The sequence of high tide levels was transformed into a sequence of (approximately) independent and identically distributed observations by first restricting attention to the 'dangerous' wintermonths (December, January and February) for homogeneity and then selecting only those high tides occurring during certain well-defined 'dangerous' windstorms for independence. An exponential distribution turned out to fit these observations well when one neglects the small ones. Once the parameters of the exponential distribution were estimated, an estimated quantile of this distribution gives an estimate for a safe level of the sea-dikes, that is a level such that the probability of having larger high tide within an arbitrary year is 1/10,000. (This is the probability set by the Dutch government.)

A recommendation was made to the government; the government fixed the new safety level for the sea-dikes and nowadays most of the dutch sea-dikes have been adapted to meet this requirement.

A few years ago the responsible government agency Rijkswaterstaat approached the Mathematical Centre (now CWI) again. It was decided to do the statistical analysis all over again, with the now-available data. This analysis is still in progress, and I would like to report on some aspects of the analysis.

A remarkable feature of the above-mentioned analysis is that little use was made of the existing literature on extreme order statistics (so-called extreme value theory). Since the fifties much progress has been made in this area both with respect to the probabilistic and to the statistical aspects. Some of these results have been obtained here at the Mathematical Centre. Before going into the details of the statistical analysis for Rijkswaterstaat, I shall first sketch some of the new (and old) developments in extreme value theory.

1. Extreme value theory

In the simplest set-up we have a sequence of independent, identically distributed (i.i.d.) random variables $X_1, X_2, \ldots$. Denote their common distribution function by F. We are interested in the distribution of

$$M_n := \max(X_1, X_2, \ldots, X_n)$$

where n is large. The distribution function of M_n is

$$P\{X_1 \leq x, \ldots, X_n \leq x\} = F^n(x) \text{ for all } x.$$

Since M_n converges to $x^* \leq \infty$ as n grows large where $x^* = \sup\{x | F(x) < 1\}$, it is difficult to calculate this function accurately for large n and we rely on asymptotic theory. Suppose there exist norming constants $a_n > 0$ and b_n ($n = 1, 2, \ldots$) such that

$$\lim_{n \to \infty} P\{\frac{M_n - b_n}{a_n} \leq x\} = \lim_{n \to \infty} F^n(a_n x + b_n) = G(x)$$

exists (for all continuity points of G) where G is a proper probability distribution function. Then we can work with G instead of F when n is large. By grouping the observations in blocks of equal size it can be seen easily that G must satisfy the following functional equation: there exist constants $A_n > 0$ and B_n such that

$$G^n(A_n x + B_n) = G(x) \text{ for all } n \in \mathbb{N}, x \in \mathbb{R}.$$

This equation is readily solved and leads to (apart from scale and shift constants)

$$G(x) = G_\gamma(x) = \exp-(1 + x/\gamma)^{-\gamma} \tag{1.1}$$

where γ is a real parameter and x such that $1 + x/\gamma \geq 0$. In particular for $\gamma = 0$

$$G(x) = \exp - e^{-x}.$$

Note that $x^* < \infty$ if and only if $\gamma < 0$. The class of distribution functions (1.1) is called the class of extreme value distributions.

For later use we note that

$$F^n(a_n x + b_n) \to G(x) \quad (n \to \infty)$$

is equivalent to

$$n\{1 - F(a_n x + b_n)\} \sim -n \log F(a_n x + b_n) \to -\log G(x) \quad (n \to \infty) \quad (1.2)$$

Hence

$$\frac{1 - F(a_n x + b_n)}{1 - F(b_n)} \to \frac{-\log G(x)}{-\log G(0)} \quad (n \to \infty).$$

It can be shown that this not only holds along a sequence: (1.2) is equivalent to ($x \geqslant 0$)

$$P\{\frac{X-t}{a(t)} > x | X > t\} = \frac{1 - F(t + xa(t))}{1 - F(t)} \to \frac{-\log G(x)}{-\log G(0)} \quad (1.3)$$

$$= (1 + x/\gamma)^{-\gamma}$$

($t \uparrow x^*$) where a is a positive function depending on F.

It is clear that the distribution functions $\{G_\gamma\}$ can be considered as good candidates for modelling extremes. We remark that if G_γ is the limiting distribution of M_n, then also the asymptotic joint distribution of the k largest observations (k fixed, $n \to \infty$) can be calculated as a function of γ.

2. STATISTICS OF EXTREME VALUES

Next we turn to the statistical part of extreme value theory and sketch two possible approaches. The basic set of data consists of a fixed number of observations per year over a large number of years. The problem is to find an approximation of the right tail (rare events) of the distribution of the yearly maximum.

A traditional approach is to form the sequence of yearly maxima and assume that this is a sequence of i.i.d. observations from *exactly* one of the distribution G_γ. One then estimates γ and the sequences $a_n > 0$ and b_n from these yearly maxima in some way. An extension of this approach is to use the m largest observations per year and then fit the asymptotic joint distribution (m fixed, $n \to \infty$) mentioned before.

A setback of the method is that, just by chance, extreme observations in one year can be significantly higher than in other years so that some larger observations are not used at all.

A different approach appears more natural: just pick the m largest observations of the *entire sequence* (so do not subdivide into years). Now we must choose m much larger. In fact we can assume (for the asymptotic theory) that m tends to infinity with n but in such way that $m/n \to 0$. The latter condition is necessary since otherwise the inference is no longer on the right tail of the distribution. Obviously the asymptotic theory for an unbounded number of high order statistics is not straightforward from the results mentioned in the previous section.

A slight variant of this approach however is conceptually much simpler: Fix a level L_n depending on n ($L_n \uparrow x^*$ as $n \to \infty$) and retain only those observations that exceed L_n. The number N_n of those 'exceedances' is then random ($\in$ Binomial $(n, 1-F(L_n))$) and given their number the exceedances are i.i.d. with distribution function

$$1 - \frac{1-F(x)}{1-F(L_n)} \text{ for } x > L_n.$$

The distribution of an exceedance E satisfies by (1.3)

$$P\{\frac{E-L_n}{a(L_n)} > x\} = \frac{1-F(L_n + xa(L_n))}{1-F(L_n)} \to (1 + x/\gamma)^{-\gamma}$$

for $n \to \infty$. Thus the normed exceedances have approximately the distribution $(1 + x/\gamma)^{-\gamma}$.

Next we discuss the problem of how to estimate the main parameter γ. Let $X_{(1)} \geq X_{(2)} \geq \ldots \geq X_{(n)}$ be the order statistics from $X_1, X_2, \ldots, X_n$ counted from above (note that we suppressed the extra index n in the notation). Let $m(n)$ be a sequence of integers satisfying $m(n) \to \infty, m(n)/n \to 0$ ($n \to \infty$). We claim that

$$\log \frac{X_{(m)} - X_{(2m)}}{X_{(2m)} - X_{(4m)}} / \log 2 \to -\gamma^{-1} \quad (n \to \infty) \text{ a.s.} \tag{2.1}$$

so that a strongly consistent estimate of γ is obtained (originally proposed by PICKANDS [3]).

The proof is a result of the following two lemma's.

LEMMA 1. *Suppose* $F(x) = 1 - e^{-x}, x > 0$ *(standard exponential distribution). Then*

$$\lim_{n \to \infty} X_{(m)} - X_{(2m)} = \log 2 \text{ a.s.}$$

PROOF. We use the following representation for exponential order statistics usually referred to as Rényi's representation: there exist i.i.d. random variables $Z_1, Z_2, \ldots$ with a standard exponential distribution, such that

$$\{X_{(m)} - X_{(m+1)}\}_{m=1}^{n-1} \stackrel{d}{=} \{Z_m/m\}_{m=1}^{n-1}.$$

This gives

$$X_{(m)} - X_{(2m)} \stackrel{d}{=} \sum_{i=m+1}^{2m} Z_i/i$$

(note that the dependence on n has disappeared).

The rest of the proof follows closely that of the classical strong law of large numbers.

Write $Q_m := \sum_{i=m+1}^{2m} (Z_i - 1)/i = \sum_{i=m+1}^{2m} Z_i/i - \log 2 + o(1)$ $(m \to \infty)$. Now $\{Q_{2^k}\}_{k=0}^{\infty}$ are independent random variables and the Borel-Cantelli lemma implies that $Q_{2^k} \to 0$ $(k \to \infty)$ a.s. Using Kolmogorov's inequality and the Borel-Cantelli lemma again one then sees that

$$\sup_{2^k \leqslant i \leqslant 2^{k+1}} (Q_i - Q_{2^k}) \to 0 \quad (k \to \infty) \text{ a.s.} \quad \square$$

LEMMA 2. *Suppose $F^n(a_n x + b_n) \to G_\gamma(x)$ for all x $(n \to \infty)$. Define $U := (\frac{1}{1-F})^{\leftarrow}$ (inverse function). Then for all $x, y > 0$, $y \neq 1$*

$$\lim_{t \to \infty} \frac{U(tx) - U(t)}{U(ty) - U(t)} = \frac{x^{-1/\gamma} - 1}{y^{-1/\gamma} - 1}.$$

PROOF. We have seen ((1.3)) that for $x \in \mathbb{R}$

$$H_t(x) := \frac{1 - F(t + xa(t))}{1 - F(t)} \to (1 + \frac{x}{\gamma})^{-\gamma} \quad (t \uparrow x^*).$$

This is a family of monotone functions converging to a continuous function. Then the inverse functions also converge:

$$H_t^{\leftarrow}(y) \to \gamma(y^{-1/\gamma} - 1) \quad (t \to \infty).$$

After working through the inversion of $H_t(x)$ one gets the stated limit relation. $\square$

PROOF OF (2.1). We are dealing with i.i.d. random variables from some distribution function F. Define as before $U := (\frac{1}{1-F})^{\leftarrow}$. Let $A_1, A_2, \ldots, A_n$ be i.i.d. exponential and $\{A_{(m)}\}$ their descending order statistics. Then $\{X_i\}_{i=1}^n = \{U(e^{A_i})\}_{i=1}^n$ are i.i.d. with distribution function F and $\{X_{(m)}\}_{m=1}^n = \{U(e^{A_{(m)}})\}_{m=1}^n$ their descending order statistics.

Now

$$\frac{X_{(m)} - X_{(2m)}}{X_{(2m)} - X_{(4m)}} = \frac{U(e^{A_{(m)}}) - U(e^{A_{(2m)}})}{U(e^{A_{(2m)}}) - U(e^{A_{(4m)}})}$$

$$= \frac{U(e^{A_{(m)} - A_{(4m)}} \cdot e^{A_{(4m)}}) - U(e^{A_{(2m)} - A_{(4m)}} \cdot e^{A_{(4m)}})}{U(e^{A_{(m)} - A_{(4m)}} \cdot e^{A_{(4m)}})}$$

$$\to \frac{\{4^{-1/\gamma} - 1\} - \{2^{-1/\gamma} - 1\}}{\{2^{-1/\gamma} - 1\}} = 2^{-1/\gamma} \quad (n \to \infty) \text{ a.s.}$$

by Lemma's 1 and 2. $\square$

A problem here is of course how to choose m in an optimal way: if m grows too fast, then the convergence of $X_{(4m)}$ to x^* goes too slowly, if m grows too slowly, then (intuitively) the accuracy of the estimate is too low since too few observations are used.

We wish to approximate $F^n(x)$ by $G_\gamma(\frac{x-b_n}{a_n})$, so apart from γ also the parameters $a_n > 0$ and b_n have to be estimated. We do not go into that problem here.

3. Peculiarities of the Rijkswaterstaat problem

We wish to apply the above theory in order to find a safe level for the Dutch sea-dikes.

An obvious problem here is that the observations (observed high tide levels) are not independent and also there is a seasonal variation, so that the marginal distribution function changes over time. The latter difficulty is approximately solved by restricting attention to the data during a period in the winter e.g. the winter months December, January and February (the stormy period).

The question of non-independence seems more serious. The procedure adopted in the Delta-report of selecting 'dangerous' windstorms by meteorological criteria and retaining only the high tide levels during those windstorms was not followed because of its complexity and subjectivity. Technically speaking the set of observations can be considered as a sequence of m-dependent random variables with m quite large i.e. observations at time points at least m units apart, are independent. It has been proved (Watson, [4]) that in this case the results for the i.i.d. case go through i.e. the limiting distributions are the same but the normalizing constants must be chosen differently, namely as if the number of observations is less than the actual number. A much more general theory along these lines has been developed by Leadbetter (the most relevant reference in this context being his 1983 paper [2]).

In Leadbetter's set-up an extra parameter is introduced called the extremal index $\theta \in [0,1]$. If $\theta = 1$, the theory for independent observations goes through without changes. If $0 < \theta < 1$, the exceedances $E_1, E_2, \ldots$ mentioned in Section 2 tend to occur in clusters (in our case during severe windstorms). The parameter $1/\theta$ basically gives the average size of a cluster.

We adopted the following method of estimating θ. The largest observation was selected and all observations that occurred within s time units before or after this observation were deleted. The largest observation from the remaining ones was selected and all observations occurring within s time units before or after this observation were deleted. And so on until all observations above a certain minimum level were either selected or deleted. The proportion of selected observations was used as an estimate of θ. The choice of s and the minimal level will not be discussed here.

Another aspect of some interest here is what to take as basic observations. Apart from the sequence of observed high tide levels there is a sequence of predicted high tide levels (so-called astronomical levels) that take into account the movements of the heavenly bodies. For several reasons it is preferable to work with 'increments' that is (roughly) the difference between actual and predicted high tide levels.

After doing the analysis based on the increments one then has to translate the conclusions for the increments into conclusions for the actual data. This can be done in the following way. Since every combination of increment and astronomical level is equally probable and since there are only few exceedances, one can consider the astronomical level as random as well. There is evidence that the increment and the astronomical level can be considered as independent random variables. We then have the problem of translating extremal results for two sequences of random variables $X_1, X_2,...$ (the increments) and $Y_1, Y_2,...$ (the astronomical levels) into extremal results for the sequence of sums $X_1 + Y_1, X_2 + Y_2,...$.

The difficulty of this problem depends on the value γ_1 and γ_2 of the two extreme value distributions involved.

Let us consider the case $\gamma_1, \gamma_2 < 0$ since that is what seems to happen in the Rijkswaterstaat problem. Then both distributions have finite upper bounds. Let us suppose that these bounds are zero. Then $F_1(0) = F_2(0) = 1$ and $F_i(x) < 1$, $x < 0$ ($i = 1, 2$). The conditions for convergence imply that F_1 and F_2 are regularly varying at 0 - i.e.

$$\lim_{t \downarrow 0} \frac{1 - F_i(-tx)}{1 - F_i(-t)} = x^{-\gamma_i} \quad (x > 0). \tag{3.1}$$

Karamata's Tauberian theorem for Laplace transforms says that then

$$\frac{\int_{-\infty}^{0} e^{s/t} dF_i(s)}{1 - F_i(-t)} = \int_{0}^{\infty} \frac{1 - F_i(-xt)}{1 - F_i(-t)} e^{-x} dx \to \Gamma(1 - \gamma_i) \quad (t \downarrow 0).$$

Denote the convolution of F_1 and F_2 by $F^{(2)}$. It follows that

$$\frac{\int_{-\infty}^{0} e^{s/t} dF^{(2)}(s)}{(1 - F_1(-t))(1 - F_2(-t))} = \frac{\int_{-\infty}^{0} e^{s/t} dF_1(s)}{1 - F_1(-t)} \cdot \frac{\int_{-\infty}^{0} e^{s/t} dF_2(s)}{1 - F_2(-t)}$$

$$\to \Gamma(1 - \gamma_1) \cdot \Gamma(1 - \gamma_2).$$

Hence (from (3.1)) the function $\hat{F}^{(2)}(t) = \int_{-\infty}^{0} e^{st} dF^{(2)}(s)$ is regularly varying at ∞ so that, again Karamata's Tauberian theorem

$$\Gamma(1 - \gamma_1 - \gamma_2)\{1 - F^{(2)}(-t)\} \sim \{1 - \hat{F}^{(2)}(1/t)\} \sim$$
$$\sim \Gamma(1 - \gamma_1) \cdot \Gamma(1 - \gamma_2)\{1 - F_1(-t)\}\{1 - F_2(-t)\} \quad (t \downarrow 0).$$

That means that the maximum of the sums $X_1 + Y_1, X_2 + Y_2, \ldots, X_n + Y_n$ behaves like the maximum of $\min(X_1, Y_1)$, $\min(X_2, Y_2)$, $\ldots, \min(X_n, Y_n)$! In the latter case combination of the asymptotic results is easy.

We remark that the corresponding problem in case $\gamma_1 = \gamma_2 = 0$ represents a well-known unsolved problem in extreme value theory.

In conclusion: the main difference between the current analysis and the analysis in the 1950's (Delta-report) is the introduction of the parameter γ. The preliminary conclusion is that with the introduction of γ the point estimate becomes somewhat lower but the confidence interval becomes wider.

4. Multidimensional extremes

One of the reasons for redoing the Delta-report was the following: In the 1950's the analysis was done mainly with the data obtained in the Rotterdam area. For other places along the coast just an adapted version of the results for the Rotterdam area was used. In the current analysis a separate treatment of the different observation stations is desirable. We can go further and consider different observation stations simultaneously. I shall give a sketch of the available probabilistic and statistical theory. In order to facilitate the exposition I only consider the 2-dimensional case.

Suppose $(X_1, Y_1), (X_2, Y_2), \ldots$ are i.i.d. observations from some distribution function $F(x,y)$.

The distribution function of $(\max_{1 \leq i \leq n} X_i, \max_{1 \leq i \leq n} Y_i)$ is $F^n(x,y)$. Suppose for simplicity that the marginal distributions of X_1 and that of Y_1 are standard exponential (this can be achieved by preliminary transformation). Suppose

$$F^n(x + \log n, y + \log n) \to G(x,y) \quad (n \to \infty),$$

a proper distribution function. Note that $\log n$ is the proper normalization for convergence of the marginals. Then as before $(n \to \infty)$

$$n\{1 - F(x + \log n, y + \log n)\} \sim -n \log F(x + \log n, y + \log n)$$
$$\to -\log G(x,y).$$

The continuous version here is

$$\lim_{t \to \infty} \frac{1 - F(t+x, t+y)}{1 - F(t,t)} = -\log G(x,y) \text{ for all continuity points } (x,y) \text{ of } G. \quad (4.1)$$

Since the lefthand side represents the measure of a set

$$A_{x,y} := \{(s,t) | s \leq x, s \leq y\}^c,$$

this must also be true for the righthand side, i.e. there is a measure v such that for all x, y

$$G(x,y) = \exp - v(A_{x,y}).$$

Moreover from the stability relation (analogous to the 1-dimensional case)

$$G^n(x + \log n, y + \log n) = G(x,y) \text{ for all } n \in \mathbb{N}, x \in \mathbb{R}$$

we get

$$n \cdot v(A + \log n) = v(A)$$

for every Borel set $A \in \mathbb{R}^2$ and $n \in \mathbb{N}$ or, more generally

$$e^u \cdot v(A+u) = v(A)$$

for every Borel set A and $u \in \mathbb{R}$. It follows that

$$v\{(s,t)|s+t>w, \ s-t \in B\} = \pi(B) \cdot e^{-w} \quad (4.2)$$

where B is a Borel set of $\mathbb{R}$ and π a measure on $[-\infty, +\infty]$ with $\int_{[-\infty,\infty]} e^{|u|} \pi(du) < \infty$. The distribution functions G are thus parametrized by a collection of (say) probability measures:

$$\begin{aligned}
G(x,y) &= v\{(u,v)|u \leqslant x, \ v \leqslant y\}^c \\
&= v\{(u,v)|(u+v)+(u-v) \leqslant 2x, (u+v)-(u-v) \leqslant 2y\}^c \\
&= v\{(u,v)|u+v > \min(2x-(u-v), 2y+(u-v))\} \\
&= \exp - \int_{[-\infty,\infty]} e^{-\min(2x-t, 2y+t)} \pi(dt) \\
&= \exp - \int_{[-\infty,\infty]} \max(e^{-2x+t}, e^{-2y-t}) \pi(dt).
\end{aligned}$$

(DE HAAN, RESNICK, [1]).

The question is how to estimate π. As in the one-dimensional situation one only considers 'high' observations, since it follows from (4.1) that for all $x,y \in \mathbb{R}$

$$P\{X_1 - t, Y_1 - t) \in A_{x,y}|(X_1, Y_1) \in A_{t,t}\} \rightarrow$$
$$-\log G(x,y) = v(A_{x,y}) \quad (t \rightarrow \infty).$$

One can prove that also the following variant holds: for each Borel set C

$$P\{(X_1 - t, Y_1 - t) \in C | X_1 + Y_1 > t\} \rightarrow v(C) \quad (t \rightarrow \infty)$$

hence, in particular (cf. (4.2))

$$P\{X_1 - Y_1 \in B | X_1 + Y_1 > t\} \rightarrow v\{s,t)|s-t \in B\} = \pi(B)$$

for each Borel set B.

This shows how one can estimate π: Consider only those observations $\{(X_{i_k}, Y_{i_k})\}_{k=1}^{N_n}$ for which the sum of the components exceeds a certain level L_n i.e. $X_{i_k} + Y_{i_k} > L_n$ for all k where $\lim_{n \rightarrow \infty} L_n = \infty$. The empirical distribution function of $X_{i_1} - Y_{i_1}, \ldots, X_{i_N} - Y_{i_N}$ is the required estimate of the probability measure π.

This procedure has not yet been applied to the rijkswaterstaat data.

ACKNOWLEDGEMENT
The sea-dike project is a joint effort of J.W. van der Made, J. van Malde and J.G. de Ronde (Rijkswaterstaat); H. Daan (Royal Meteorological Institute); A.L.M. Dekkers, L. de Haan and R. Helmers (CWI). The responsibility for the

present text is completely the author's. A discussion with H. Berbee on Lemma 1 is grateful acknowledged.

REFERENCES
1. L. DE HAAN, S.I. RESNICK (1977). Limit theory for multivariate sample extremes. *Z. Wahrsch. verw. Gebiete 40,* 317-337.
2. M.R. LEADBETTER (1983). Extremes and local dependence in stationary sequences. *Z. Wahrsch. verw. Gebiete 65,* 291-306.
3. J. PICKANDS III (1975). Statistical inference using extreme order statistics. *Ann. Statist. 3,* 119-131.
4. G.S. WATSON (1954). Extreme values in samples from m-dependent stationary stochastic processes. *Ann. Math. Statist. 25,* 798-800.

Process Algebra: Specification and Verification in Bisimulation Semantics

J.A. Bergstra

University of Amsterdam, Department of Computer Science
P.O. Box 19268, 1000 GG Amsterdam, The Netherlands
and
State University of Utrecht, Department of Philosophy
P.O. Box 80010, 3508 TA Utrecht, The Netherlands

J.W. Klop

Centre for Mathematics and Computer Science
P.O. Box 4079, 1009 AB Amsterdam, The Netherlands

This paper addresses itself primarily to readers who have not had much exposure to algebraic approaches to concurrency, or as we will call it, process algebra. We will describe an algebraic framework called $ACP_\tau^\#$ (Algebra of Communicating Processes with abstraction and additional features), which is suitable for both specification and verification of communicating processes. Except in two instances we give no proofs; but there are many references to the places where these can be found. One instance where we do give a proof is the verification of the Alternating Bit Protocol. Here the point is that an algebraic proof can be given. The formal system $ACP_\tau^\#$ is, at least theoretically, very close to a universal system for process specification: every finitely branching computable process, can be finitely specified. In practice one needs additional operators for specifications; some of these are briefly discussed in a final section.

Our presentation will concentrate on process algebra as it has been developed since 1982 at the Centre for Mathematics and Computer Science, since 1985 in cooperation with the University of Amsterdam and the University of Utrecht. This means that we make no attempt to give a survey of related approaches though there will be references to some of the main ones.

This paper is not intended to give a survey of the whole area of activities in process algebra. Specifically, we will restrict ourselves to that side of the spectrum of process semantics which was initiated by MILNER [30] and which is

---

1. This research was partially sponsored by ESPRIT project nr. 432, Meteor.

called 'bisimulation semantics'. Thus, the important aspect of process algebra in which a unification and classification is sought for various algebraical approaches to process semantics ('comparative concurrency semantics') is not represented here. From the point of view of process specification and verification this restriction is justified: at present the specification and verification facilities are, at least in the setting of ACP, most highly developed in bisimulation semantics, in any case more than in the ACP treatment of e.g. failure semantics.

ACKNOWLEDGEMENT. We thank J. Heering and J.C.M. Baeten for suggesting many improvements.

1. THE BASIC CONSTRUCTORS

The processes that we will consider are capable of performing atomic steps or actions $a,b,c, \ldots$, with the idealization that these actions are events without positive duration in time; it takes only one moment to execute an action. The actions are combined into composite processes by the operations $+$ and $\cdot$, with the interpretation that $(a+b)\cdot c$ is the process that first chooses between executing a or b and, second, performs the action c after which it is finished. (We will often suppress the dot and write $(a+b)c$.) These operations, 'alternative composition' and 'sequential composition' (or just sum and product), are the basic constructors of processes. Since time has a direction, multiplication is not commutative; but addition is, and in fact it is stipulated that the options (summands) possible at some stage of the process form a *set*. Formally, we will require that processes $x,y,\ldots$ satisfy the following axioms:

BPA
$x+y=y+x$
$(x+y)+z=x+(y+z)$
$x+x=x$
$(x+y)z=xz+yz$
$(xy)z=x(yz)$

TABLE 1

Thus far we used 'process algebra' in the generic sense of denoting the area of algebraic approaches to concurrency, but we will also adopt the following technical meaning for it: any model of these axioms will be a *process algebra*. The simplest process algebra, then, is the term model of BPA (Basic Process Algebra), whose elements are BPA-expressions (built from the atoms $a,b,c,\ldots$ by means of the basic constructors) modulo the equality generated by the axioms. This process algebra contains only finite processes; things get more lively if we admit recursion enabling us to define infinite processes. Even at this stage one can define, recursively, interesting processes:

COUNTER
$X = (zero + up.Y).X$
$Y = down + up.Y.Y$

TABLE 2

where 'zero' is the action that asserts that the counter has value 0, and 'up' and 'down' are the actions of incrementing resp. decrementing the counter by one unit. The process COUNTER is now represented by X; Y is an auxiliary process. COUNTER is a 'perpetual' process, that is, all its execution traces are infinite. Such a trace is e.g. zero-zero-up-down-zero-up-up-up-.... A question of mathematical interest only is: can COUNTER be defined in a single equation, without auxiliary processes? The negative answer is an immediate consequence of the following fact:

THEOREM 1. *Let a system* $\{X_i = T(X_1, \ldots, X_n) | \; i = 1, \ldots, n\}$ *of guarded fixed point equations over BPA be given. Suppose the solutions X_i are all perpetual. Then they are regular.*

Two concepts in this statement need explanation: a fixed point equation, like $X = (zero + up.Y).X$ is *guarded* if every occurrence of a recursion variable in the right hand side is preceded ('guarded') by an occurrence of an action. For instance, the occurrence of X in the RHS of $X = (zero + up.Y).X$ is guarded since, when this X is accessed, one has to pass either the guard zero or the guard up. A non-example: the equation $X = X + a.X$ is not guarded. Furthermore, a process is *regular* if it has only finitely many 'states'; clearly, COUNTER is not regular since it has just as many states as there are natural numbers. Let us mention one other property of processes which have a finite recursive specification (by means of guarded recursion equations) in BPA: such processes are *uniformly finitely branching*. A process is finitely branching if in each of its states it can take steps (and thereby transform itself) to only finitely many subprocesses; for instance, the process defined by $X = (a + b + c)X$ has in each state branching degree 3. 'Uniformly' means that there is uniform bound on the branching degrees throughout the process.

In fact, a more careful treatment is necessary to define concepts like 'branching degree' rigorously. For, clearly, the branching degree of $a + a$ ought to be the same as that of the process 'a', since $a + a = a$. And the process $X = aX$ will be the same as the process $X = aaX$; in turn these will be identified with the process $X = aX + aaX$. In the sequel we will discuss the semantic criterion by means of which these processes are identified ('bisimilarity'). MILNER [31] has found a simple axiom system (extending BPA) which is able to deal with recursion and which is complete for regular processes with respect to 'bisimilarity'.

Before proceeding to the next section, let us assure the reader that the omission of the other distributive law, $z(x+y)=zx+zy$, is intentional. The reason will become clear after the introduction of 'deadlock'.

2. Deadlock

A vital element in the present set-up of process algebra is the process δ, signifying 'deadlock'. The process ab performs its two steps and then stops, silently and happily; but the process $ab\delta$ deadlocks (with a crunching sound, one may imagine) after the a- and b-action: it wants to do a proper action but it cannot. So δ is the acknowledgement of stagnation. With this in mind, the axioms to which δ is subject, should be clear:

DEADLOCK
$\delta + x = x$
$\delta . x = \delta$

Table 3

(In fact, it can be argued that 'deadlock' is not the most appropriate name for the process constant δ. In the sequel we will encounter a process which can more rightfully claim this name: $\tau\delta$, where τ is the silent step. We will stick to the present terminology, however.)

The axiom system of BPA (Table 1) together with the present axioms for δ is called BPA$_\delta$. Now suppose that the distributive law $z(x+y)=zx+zy$ is added to BPA$_\delta$. Then: $ab=a(b+\delta)=ab+a\delta$. This means that a process without deadlock possibility is equal to one without; and that conflicts with our intention to model also deadlock behaviour of processes.

3. Interleaving, or free merge

If x, y are processes, their 'parallel composition' $x\|y$ is the process that first chooses whether to do a step in x or in y, and proceeds as the parallel composition of the remainders of x, y. In other words, the steps of x, y are interleaved. Using an auxiliary operator $\|\!\_$ (with the interpretation that $x\|\!\_y$ is like $x\|y$ but with the commitment of choosing the initial step from x) the operation $\|$ can be succintly defined by the axioms:

FREE MERGE
$x\|y = x\|\!\_y + y\|\!\_x$
$ax\|\!\_y = a(x\|y)$
$a\|\!\_y = ay$
$(x+y)\|\!\_z = x\|\!\_z + y\|\!\_z$

Table 4

One can show that an equivalent axiomatization of $\parallel$ without an auxiliary operator like $\parallel\!\!\_$, would require infinitely many axioms.

The system of nine axioms consisting of BPA and the four axioms for free merge will be called PA. Moreover, if the axioms for δ are added, the result will be PA_δ. The operators $\parallel$ and $\parallel\!\!\_$ will also be called *merge* and *left-merge* respectively.

An example of a process recursively defined in PA, is: $X = a(b\parallel X)$. It turns out that this process can already be defined in BPA, by the two fixed point equations $X = aYX$, $Y = b + aYY$. (This is a simplified version of the counter in Table 2, without the action zero.) To see that both ways of defining X yield the same process, one may 'unwind' according to the given equations: $X = a(b\parallel X) = a(b\parallel\!\!\_ X + X\parallel\!\!\_ b) = a(bX + a(b\parallel X)\parallel\!\!\_ b) = a(bX + a((b\parallel X)\parallel b)) = a(bX + a...)$, while on the other hand $X = aYX = a(b + aYY)X = a(bX + aYYX) = a(bX + a...)$; so at least up to level 2 the processes are equal. In fact they can be proved equal up to each finite level. Later on, we will introduce an infinitary proof rule enabling us to infer that, therefore, the processes are equal.

So, is the defining power (or expressibility) of PA greater than that of BPA? Indeed it is, as is shown by the following process:

BAG
$X = in(0)(out(0)\parallel X) + in(1)(out(1)\parallel X)$

TABLE 5

This equation describes the process behaviour of a 'bag' or 'multiset' that may contain finitely many instances of data 0, 1. The actions $in(0)$, $out(0)$ are: putting a 0 in the bag resp. getting a 0 from the bag, and likewise for 1. This process does not have a finite specification in BPA, that is, a finite specification without merge ($\parallel$). We conclude this section about PA by mentioning the following fact:

THEOREM 2. *Every process which is recursively defined in PA and has an infinite trace, has an eventually periodic trace.*

4. FIXED POINTS

We have already alluded to the existence of infinite processes; this raises the question how one can actually construct process algebras (for BPA or PA) containing infinite processes in addition to finite ones. Such models can be obtained as:
(1) projective limits ([14,15]);
(2) complete metrical spaces, as in the work of DE BAKKER and ZUCKER [6,7];
(3) quotients of graph domains (a graph domain is a set of process graphs or transition diagrams), as in MILNER [30];

(4) the 'explicit' models of HOARE [25];
(5) ultraproducts of finite models (KRANAKIS [28]).

In Section 13 we will discuss a model as in (3). As to (5), these models are only of theoretical interest: models thus obtained contain 'weird' processes such as $x = \sqrt{a^\omega}$, a process satisfying $x^2 = a^\omega = a.a.a...$ while $x \neq x^2$.

Here, we look at (2). First, define the projection operators $\pi_n (n \geq 1)$, cutting off a process at level n:

PROJECTION
$\pi_1(ax) = a$
$\pi_{n+1}(ax) = a\pi_n(x)$
$\pi_n(a) = a$
$\pi_n(x+y) = \pi_n(x) + \pi_n(y)$

TABLE 6

E.g., for X defining BAG:

$$\pi_2(X) = in(0)(out(0) + in(0) + in(1)) + in(1)(out(1) + in(0) + in(1)).$$

By means of these projections a distance between processes x, y can be defined: $d(x,y) = 2^{-n}$ where n is the least natural number such that $\pi_n(x) \neq \pi_n(y)$, and $d(x,y) = 0$ if there is no such n. If the term model of BPA (or PA) as in Section 1 is equipped with this distance function, the result is an ultrametrical space. By metrical completion we obtain a model of BPA (resp. PA) in which all systems of guarded recursion equations have a unique solution. Call this model the *standard model*. In fact, the guardedness condition is exactly what is needed to associate a contracting operator on the complete metrical space with a guarded recursion equation. (E.g. to the recursion equation $X = aX$ the contracting function $f(x) = ax$ is associated; indeed $d(f(x), f(y)) \leq d(x,y)/2$.) The contraction theorem of Banach then proves the existence of a unique fixed point. This model construction has been employed in various settings by DE BAKKER and ZUCKER [6,7], who posed the question whether *unguarded* fixed point equations, such as $X = aX + X$ or $Y = (aY \| Y) + b$, always have a solution in the standard model as well. This turns out to be the case:

THEOREM 3 ([10]). *Let q be an arbitrary process in the standard model, and let $X = s(X)$ be a recursion equation in the signature of PA. Then the sequence q, $s(q), s(s(q)), s(s(s(q))),...$ converges to a solution $q^* = s(q^*)$.*

In general, the fixed points $q^* = s(q^*)$ are not unique. The proof in [10] is combinatorial in nature; it is not at all clear whether this convergence result can be obtained by the 'usual' convergence proof methods, such as invoking

Banach's fixed point theorem or (in a complete partial order setting) the Knaster-Tarski fixed point theorem. In KRANAKIS [29] the present theorem is extended to the case where $s(X)$ may contain parameters.

5. COMMUNICATION

So far, the parallel composition or merge ($\|$) did not involve communication in the process $x\|y$: x and y are 'freely' merged. However, some actions in one process may need an action in another process for an actual execution, like the act of shaking hands requires simultaneous acts of two persons. In fact, 'hand shaking' is the paradigm for the type of communication which we will introduce now. If $A = \{a,b,c,\ldots,\delta\}$ is the action alphabet, let us adopt a binary communication function $|:A \times A \to A$ satisfying

COMMUNICATION FUNCTION
$a\|b = b\|a$
$(a\|b)\|c = a\|(b\|c)$
$\delta\|a = \delta$

TABLE 7

(Here a,b vary over A, including δ.) We can now specify *merge with communication;* we use the same notation $\|$ as for the free merge, since in fact free merge is an instance of merge with communication (by choosing the communication function trivial, i.e. $a\|b = \delta$ for all a,b). There are now two auxiliary operators, allowing a finite axiomatisation: left-merge ($\mathop{\|\!\_}$) as before and $|$ (communication merge or 'bar'), which is an extension of the communication function to all processes, not only the atoms. The axioms for $\|$ and its auxiliary operators are:

MERGE WITH COMMUNICATION
$x\|y = x \mathop{\|\!\_} y + y \mathop{\|\!\_} x + x\|y$
$ax \mathop{\|\!\_} y = a(x\|y)$
$a \mathop{\|\!\_} y = ay$
$(x+y) \mathop{\|\!\_} z = x \mathop{\|\!\_} z + y \mathop{\|\!\_} z$
$ax\|b = (a\|b)x$
$a\|bx = (a\|b)x$
$ax\|by = (a\|b)(x\|y)$
$(x+y)\|z = x\|z + y\|z$
$x\|(y+z) = x\|y + x\|z$

TABLE 8

We also need the so-called *encapsulation* operators $\partial_H (H \subseteq A)$ for removing unsuccessful attempts at communication:

ENCAPSULATION
$\partial_H(a) = a$ if $a \notin H$
$\partial_H(a) = \delta$ if $a \in H$
$\partial_H(x+y) = \partial_H(x) + \partial_H(y)$
$\partial_H(xy) = \partial_H(x).\partial_H(y)$

TABLE 9

The axioms for BPA, DEADLOCK together with the present ones constitute the axiom system ACP (Algebra of Communicating Processes). Typically, a system of communicating processes $x_1, \ldots, x_n$ is now represented in ACP by the expression $\partial_H(x_1 \| \ldots \| x_n)$. Prefixing the encapsulation operator says that the system $x_1, \ldots, x_n$ is to be perceived as a separate unit w.r.t. the communication actions mentioned in H; no communications between actions in H with an environment are expected or intended. A useful theorem to break down such expressions is the *Expansion Theorem* which holds under the assumption of the *handshaking axiom* $x|y|z = \delta$. This axiom says that all communications are binary. (In fact we have to require associativity of '|' first - see Table 10.)

THEOREM 4 (Expansion Theorem).

$$x_1 \| \ldots \| x_k = \Sigma_i x_i \mathbin{\|\mkern-9mu\relbar} X^i_k + \Sigma_{i \neq j}(x_i | x_j) \mathbin{\|\mkern-9mu\relbar} X^{i,j}_k$$

Here X^i_k denotes the merge of $x_1, \ldots, x_k$ except x_i, and $X^{i,j}_k$ denotes the same merge except $x_i, x_j (k \geq 3)$. In order to prove the expansion theorem, one first proves by simultaneous induction on term complexity that for all closed ACP-terms (i.e. ACP-terms without free variables) the following holds:

AXIOMS OF STANDARD CONCURRENCY
$(x \mathbin{\|\mkern-9mu\relbar} y) \mathbin{\|\mkern-9mu\relbar} z = x \mathbin{\|\mkern-9mu\relbar} (y \| z)$
$(x|y) \mathbin{\|\mkern-9mu\relbar} z = x|(y \mathbin{\|\mkern-9mu\relbar} z)$
$x|y = y|x$
$x \| y = y \| x$
$x|(y|z) = (x|y)|z$
$x \| (y \| z) = (x \| y) \| z$

TABLE 10

(As in Section 4 we can construct the 'standard' model for ACP; in this model the above axioms are valid. We will return to the existence and construction of models later.)

What about the defining power of ACP? The following is an example of a process p, recursively defined in ACP, but not definable in PA: let the alphabet be $\{a,b,c,d,\delta\}$ and let the communication function be given by $c|c=a$, $d|d=b$, and all other communications equal to δ. Let $H=\{c,d\}$.

$$\boxed{\begin{array}{l} X=cXc+d \\ Y=dXY \\ Z=dXcZ \\ p=\partial_H(dcY\|Z) \end{array}}$$

Then $p=ba(ba^2)^2(ba^3)^2(ba^4)^2\ldots$. Indeed, using the axioms in ACP and putting $p_n=\partial_H(dc^nY\|Z)$ for $n\geq 1$, one proves that $p_n=ba^nba^{n+1}p_{n+1}$ (see [11]). By Theorem 2 in Section 3, p is not definable in PA, since the one infinite trace of p is not eventually periodic.

We will often adopt the following special format for the communication function, called *read-write communication*. Let a finite set D of *data* d and a set $\{1,\ldots,p\}$ of *ports* be given. Then the alphabet consists of *read* actions $ri(d)$ and *write* actions $wi(d)$, for $i=1,\ldots,p$ and $d\in D$. The interpretation is: read datum d at port i, resp. write datum d at port i. Furthermore, the alphabet contains actions $ci(d)$ for $i=1,\ldots,p$ and $d\in D$, with interpretation: *communicate d at i*. These actions will be called *transactions*. The only non-trivial communications (i.e. not resulting in δ) are: $wi(d)|ri(d)=ci(d)$. Instead of $wi(d)$ we will also use the notation $si(d)$ (send d along i). Note that read-write communication satisfies the hand-shaking axiom: all communications are binary.

In order to illustrate the defining power of ACP, we will now give an infinite specification of the process behaviour of a queue with input port 1 and output port 2. Here D is a finite set of data (finite since otherwise the sums in the specification below would be infinite, and we do not consider infinite expressions), D^* is the set of finite sequences σ of elements from D; the empty sequence is λ. The sequence $\sigma*\sigma'$ is the concatenation of sequences σ,σ'.

$$\boxed{\begin{array}{c} \text{QUEUE} \\ Q=Q_\lambda=\Sigma_{d\in D}r1(d).Q_d \\ Q_{\sigma*d}=s2(d).Q_\sigma+\Sigma_{e\in D}r1(e).Q_{e*\sigma*d} \quad \text{(for all } d\in D \text{ and } \sigma\in D^*\text{)} \end{array}}$$

TABLE 11

Note that this infinite specification uses only the signature of BPA. We have the following remarkable fact:

THEOREM 5. *Using read-write communication, the process Queue cannot be specified in ACP by finitely many recursion equations.*

For the lengthy proof see [2,19]. It should be mentioned that the process Queue can be finitely specified in ACP if the read-write restriction is dropped and n-ary communications are allowed; in the next section it is shown how this can be done. In the sequel we will present some other finite specifications of Queue using features to be introduced later.

6. Renaming

A useful 'add-on' feature is formed by the renaming operators ρ_f, where $f: A \to A$ is a function keeping δ fixed. A renaming ρ_f replaces each action 'a' in a process by $f(a)$. In fact, the encapsulation operators ∂_H are renaming operators; f maps $H \subseteq A$ to δ and fixes $A - H$ pointwise. The following axioms, where 'id' is the identity function, are obvious:

RENAMING
$\rho_f(a) = f(a)$
$\rho_f(x+y) = \rho_f(x) + \rho_f(y)$
$\rho_f(xy) = \rho_f(x).\rho_f(y)$
$\rho_{id}(x) = x$
$(\rho_f \circ \rho_g)(x) = \rho_{f \circ g}(x)$

Table 12

Again the defining power is enhanced by adding this feature. While Queue as in the previous section could not yet be finitely specified, it can now.

The actions are the $r1(d)$, $s2(d)$ as before; there are moreover 'auxiliary' actions $r3(d)$, $s3(d)$, $c3(d)$ for each datum d. Communication is given by $r3(d)|s3(d) = c3(d)$ and there are no other communications. If we let $\rho_{c3 \to s2}$ be the renaming $c3(d) \to s2(d)$ and $\rho_{s2 \to s3}: s2(d) \to s3(d)$, then for $H = \{s3(d), r3(d) | d \in D\}$ the following two guarded recursion equations give a finite specification of Queue:

QUEUE, FINITE SPECIFICATION
$Q = \Sigma_{d \in D} r1(d)(\rho_{c3 \to s2} \circ \partial_H)(\rho_{s2 \to s3}(Q) \| s2(d).Z)$
$Z = \Sigma_{d \in D} r3(d).Z$

Table 13

(This little gem was inspired by a similar specification in Hoare [24]. The present formulation is from Baeten and Bergstra [2].) The explanation that this is really Queue is as follows. We intend that Q processes data d in a queue-like manner, by performing 'input' actions $r1(d)$ and 'output' actions $s2(d)$. So $\rho_{s2 \to s3}(Q)$ processes data in queue-like manner by performing input actions $r1(d)$, output actions $s3(d)$. First consider the parallel system

$Q' = \partial_H(\rho_{s2\to s3}(Q)\|Z)$: since Z universally accepts $s3(d)$ and transforms these into $c3(d)$, this is just the queue with input $r1(d)$, output $c3(d)$. Now the process $Q^* = \partial_H(\rho_{s2\to s3}(Q)\|s2(d).Z)$ appearing in the recursion equation, is just like Q' but with the obligation to perform output action $s2(d)$ *before all output actions* $c3(d)$; this obligation is enforced since $s2(d)$ must be passed before $\rho_{s2\to s3}(Q)$ and Z can communicate and thereby create the output actions $c3(d)$. So $\rho_{c3\to s2}(Q^*) = Q_d$, the queue loaded with d, in the earlier notation used for the infinite specification of Queue (Table 11). But then $Q = \Sigma_{d\in D} r1(d).Q_d$ and this is exactly what we want.

In fact, the renamings used in this specification can be removed in favour of a more complicated communication format, as follows. Replace in the specification above $\rho_{s2\to s3}(Q)$ by $\partial_{s2}(Q\|V)$ where $V = \Sigma_d s2^*(d).V$ and $S2 = \{s2(d), s2^*(d)|d\in D\}$ with communications $s2(d)|s2^*(d) = s3(d)$ for all d. To remove the other renaming operator, put $P = \partial_H(\partial_{s2}(Q\|V)\|s2(d).Z)$, and replace $\rho_{c3\to s2}(P)$ by $\partial_{C3}(P\|W)$ where $W = \Sigma_d c3^*(d).W$ and $c3(d)|c3^*(d) = s2(d)$ for all d. However, though the renamings are removed in this way, the communication is no longer of the read-write format, or even in the hand shaking format, since we have ternary nontrivial communications $s2(d) = c3(d)|c3^*(d) = r3(d)|s3(d)|c3^*(d)$. As we already stated in the last theorem, this is unavoidable.

7. Abstraction

A fundamental issue in the design and specification of hierarchical (or modularized) systems of communicating processes is *abstraction*. Without having an abstraction mechanism enabling us to abstract from the inner workings of modules to be composed to larger systems, specification of all but very small systems would be virtually impossible. We will now extend the axiom system ACP, obtained thus far, with such an abstraction mechanism. Consider two bags B_{12}, B_{23} (cf. Section 3) with action alphabets $\{r1(d), s2(d)|d\in D\}$ resp. $\{r2(d), s3(d)|d\in D\}$. That is, B_{12} is a bag-like channel reading data d at port 1, sending them at port 2; B_{23} reads data at 2 and sends them to 3. (That the channels are bags means that, unlike the case of a queue, the order of incoming data is lost in the transmission.) Suppose the bags are connected at 2; that is, we adopt communications $s2(d)|r2(d) = c2(d)$ where $c2(d)$ is the transaction of d at 2.

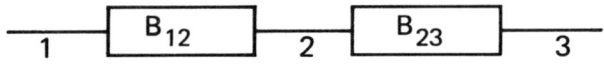

FIGURE 1

The composite system $\mathbf{B}_{13} = \partial_H(B_{12}\|B_{23})$ where $H = \{s2(d), r2(d)|d\in D\}$, should, intuitively, be again a bag between locations 2, 3. However, some (rather involved) calculations learn that $\mathbf{B}_{13} = \Sigma_{d\in D} r1(d).((c2(d)s3(d))\|\mathbf{B}_{13})$; so $\mathbf{B}_{13}$ is a 'transparant' bag: the passage of d through 2 is visible as the transaction event $c\,2(d)$.

How can we *abstract* from such internal details, if we are only interested in the external behaviour at 1, 3? The first step to obtain such an abstraction is to remove the distinctive identity of the actions to be abstracted, that is, to rename them all into one designated action which we call, after Milner, τ: the *silent* action (this is called 'pre-abstraction' in [2]). This special renaming is the *abstraction operator* τ_I, parameterized by a set of actions $I \subseteq A$ and subject to the following axioms:

ABSTRACTION
$\tau_I(\tau) = \tau$
$\tau_I(a) = a$ if $a \notin I$
$\tau_I(a) = \tau$ if $a \in I$
$\tau_I(x+y) = \tau_I(x) + \tau_I(y)$
$\tau_I(xy) = \tau_I(x).\tau_I(y)$

TABLE 14

The second step is to attempt to devise axioms for the silent step τ by means of which τ can be removed from expressions, as e.g. in the equation $a\tau b = ab$. However, it is not possible (nor desirable) to remove *all* τ's in an expression if one is interested in a faithful description of deadlock behaviour of processes. For, consider the process (expression) $a + \tau\delta$; this process can deadlock, namely if it chooses to perform the silent action. Now, if one would propose naively the equations $\tau x = x\tau = x$, then $a + \tau\delta = a + \delta = a$, and the latter process has no deadlock possibility. It turns out that one of the proposed equations, $x\tau = x$, can safely be adopted, but the other one is wrong. Fortunately, MILNER [31] has devised some simple axioms which give a complete description of the properties of the silent step (complete w.r.t a certain semantical notion of process equivalence called bisimulation, which does respect deadlock behaviour; this notion is discussed in the sequel), as follows.

SILENT STEP
$x\tau = x$
$\tau x = \tau x + x$
$a(\tau x + y) = a(\tau x + y) + ax$

TABLE 15

To return to our example of the transparant bag $\mathbf{B}_{13}$, after abstraction of the set of transactions $I = \{c\,2(d) | d \in D\}$ the result is indeed an 'ordinary' bag:

$$\tau_I(\mathbf{B}_{13}) = \tau_I(\Sigma r\, 1(d)(c\,2(d).s\,3(d) \| \mathbf{B}_{13})) =^{(*)} \Sigma r\, 1(d)(\tau.s\,3(d) \| \tau_I(\mathbf{B}_{13}))$$

$$= \Sigma(r\,1(d).\tau.s\,3(d)) \mathbin{\|\!\_} \tau_I(\mathbf{B}_{13}) = \Sigma(r\,1(d).s\,3(d)) \mathbin{\|\!\_} \tau_I(\mathbf{B}_{13})$$
$$= \Sigma r\,1(d)(s\,3(d) \| \tau_I(\mathbf{B}_{13}))$$

from which it follows that $\tau_I(\mathbf{B}_{13}) =^{(**)} B_{13}$, the bag defined by

$$B_{13} = \Sigma r\,1(d)(s\,3(d) \| B_{13}).$$

Here we were able to eliminate all silent actions, but this will not always be the case. In fact, this computation is not as straightforward as was maybe suggested: to justify the equations marked with (*) and (**) we need more powerful principles, which we will discuss now. (Specifically, in (*) an appeal to the 'alphabet calculus' below is needed and (**) requires the principle RSP, also below.)

8. Proof rules for recursive specifications

We have now presented a survey of ACP_τ; we refer to [12] for an analysis of this proof system as well as a proof that (when the hand shaking axiom is adopted) the Expansion Theorem carries over from ACP to ACP_τ unchanged. Note that ACP_τ (displayed in full in Section 11) is entirely equational. Without further proof rules it is not possible to deal (in an algebraical way) with infinite processes, obtained by recursive specifications, such as Bag; in the derivation above we tacitly used such proof rules which will be made explicit now.

(i) RDP, the Recursive Definition Principle: *Every guarded and abstraction free recursive specification has a solution.*

(ii) RSP, the Recursive Specification Principle: *Every guarded and abstraction free recursive specification has at most one solution.*

(iii) AIP, the Approximation Induction Principle: *A process is determined by its finite projections.*

In a more formal notation, AIP can be rendered as the infinitary rule

$$\frac{\forall n \quad \pi_n(x) = \pi_n(y)}{x = y}$$

As to (i), the restriction to guarded specifications is not very important (for the definition of 'guarded' see Section 1); in the process algebras that we have encountered and that satisfy RDP, also the same principle without the guardedness condition is true. More delicate is the situation in principle (ii): first, *τ-steps may not act as guards:* e.g. the recursion equation $X = \tau X + a$ has infinitely many solutions, namely $\tau(a + q)$ is a solution for arbitrary q; and second, the *recursion equations must not contain occurrences of abstraction operators τ_I*. That is, they are 'abstraction-free' (but there may be occurrences of τ in the equations). The latter restriction is in view of the fact that, surprisingly, the recursion equation $X = a.\tau_{\{a\}}(X)$ possesses infinitely many solutions, even though it looks very guarded. (The solutions are: $a.q$ where q satisfies $\tau_{\{a\}}(q) = q$.) That the presence of abstraction operators in recursive specifications causes trouble, was first noticed by Hoare [24,25].

As to (iii), we still have to define projections π_n in the presence of the τ-

action. The extra clauses are:

PROJECTION, CONTINUED
$\pi_n(\tau) = \tau$
$\pi_n(\tau x) = \tau.\pi_n(x)$

TABLE 16

So, τ-steps do not add to the depth; this is enforced by the τ-laws (since, e.g., $a\tau b = ab$ and $\tau a = \tau a + a$). Remarkably, there are infinitely many different terms t_n (that is, different in the term model of ACP$_\tau$), built from τ and a single atom a, such that t_n has depth 1, i.e. $t = \pi_1(t)$. The t_n are inductively defined as follows:
$t_0 = a$, $t_1 = \tau a$, $t_2 = \tau$, $t_3 = \tau(a + \tau)$, $t_4 = a + \tau a$, $t_{4k+i} = \tau.t_{4k+i-1}$ for $i = 1, 3$, $t_{4k+i} = t_{4k+i-3} + t_{4k+i-5}$ for $i = 0, 2$.

The unrestricted form of AIP as in (iii) will turn out to be too strong in some circumstances; it does not hold in one of the main models of ACP$_\tau$, namely the graph model which is introduced in Section 13. Therefore we also introduce the following weaker form.

(iv) AIP$^-$ (Weak Approximation Induction Principle): *Every process which has an abstraction-free guarded specification is determined by its finite projections.*

Roughly, a process which can be specified without abstraction operators is one in which there are no infinite τ-traces (and which is definable). E.g. the process X_0 defined by the infinite specification $\{X_0 = bX_1, X_{n+1} = bX_{n+2} + a^n\}$, where a^n is a.a.....a (n times), contains an infinite trace of b-actions; after abstraction w.r.t. b, the resulting process, $Y = \tau_{\{b\}}(X_0)$, has an infinite trace of τ-steps; and (at least in the main model of ACP$_\tau$ of Section 13) this Y is not definable without abstraction operators.

Even the Weak Approximation Induction Principle is rather strong. In fact a short argument shows the following:

THEOREM 6. AIP$^- \Rightarrow$ RSP.

As a rule, we will be very careful in admitting abstraction operators in recursive specifications. Yet there are processes which can be elegantly specified by using abstraction inside recursion. The following curious specification of Queue is obtained in this manner. We want to specify Q_{12}, the queue from port 1 to 2, using an auxiliary port 3 and concatenating auxiliary queues Q_{13}, Q_{32}; then we abstract from the internal transaction at port 3. Write, in an ad hoc notation, $Q_{12} = Q_{13} * Q_{32}$. Now Q_{13} can be similarly split up: $Q_{13} = Q_{12} * Q_{32}$. This gives rise to six similar equations: $Q_{ab} = Q_{ac} * Q_{cb}$ where $\{a,b,c\} = \{1,2,3\}$. (See Figure 2.)

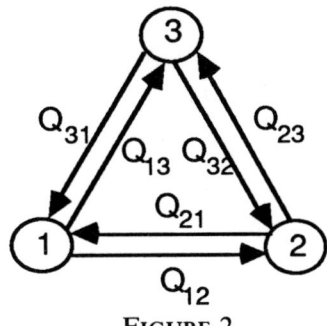

FIGURE 2

These six queues, which are merely renamings of each other, can now be specified in terms of each other as in the following table. One can prove that these recursion equations, though not abstraction-free, indeed have a unique solution.

QUEUE, FINITE SPECIFICATION WITH ABSTRACTION
$Q_{ab} = \Sigma_{d \in D} ra(d).\tau_c \circ \partial_c (Q_{ac} \| sb(d).Q_{cb})$ for $\{a,b,c\} = \{1,2,3\}$

TABLE 17

Here the usual read-write notation is used: $ri(d)$ means read d at i, $si(d)$: send d at i, communications are $ri(d)|si(d) = ci(d)$; further $\tau_i = \tau_{\{ci(d)|d \in D\}}$ and $\partial_i = \partial_{\{ri(d), si(d)|d \in D\}}$. This example shows that even with the restriction to read-write communication, ACP_τ is stronger than ACP.

9. ALPHABET CALCULUS

In computations with infinite processes one often needs information about the *alphabet* $\alpha(x)$ of a process x. E.g. if x is the process uniquely defined by the recursion equation $X = aX$, we have $\alpha(x) = \{a\}$. An example of the use of this alphabet information is given by the implication $\alpha(x) \cap H = \emptyset \Rightarrow \partial_H(x) = x$. For finite closed process expressions this fact can be proved with induction to the structure, but for infinite processes we have to require such a property axiomatically. In fact, the example will be one of the 'conditional axioms' below (conditional, in contrast with the purely equational axioms we have introduced thus far). First we have to define the alphabet:

ALPHABET
$\alpha(\delta) = \emptyset$
$\alpha(\tau) = \emptyset$
$\alpha(a) = \{a\}$
$\alpha(\tau x) = \alpha(x)$
$\alpha(ax) = \{a\} \cup \alpha(x)$
$\alpha(x + y) = \alpha(x) \cup \alpha(y)$
$\alpha(x) = \bigcup_{n \geq 1} \alpha(\pi_n(x))$
$\alpha(\tau_I(x)) = \alpha(x) - I$

TABLE 18

To appreciate the non-triviality of the concept $\alpha(x)$, let us mention that a finite specification can be given of a process for which the alphabet is uncomputable (see [3] for an example).

Now the following conditional axioms will be adopted:

CONDITIONAL AXIOMS
$\alpha(x)\|(\alpha(y) \cap H) \subseteq H \Rightarrow \partial_H(x\|y) = \partial_H(x\|\partial_H(y))$
$\alpha(x)\|(\alpha(y) \cap I) = \emptyset \Rightarrow \tau_I(x\|y) = \tau_I(x\|\tau_I(y))$
$\alpha(x) \cap H = \emptyset \Rightarrow \partial_H(x) = x$
$\alpha(x) \cap I = \emptyset \Rightarrow \tau_I(x) = x$

TABLE 19

Using these axioms, one can derive for instance the following fact: if communication is of the read-write format and I is disjoint from the set of transactions (communication results) as well as disjoint from the set of communication actions, then the abstraction τ_I distributes over merges $x\|y$.

10. KOOMEN'S FAIR ABSTRACTION RULE

Suppose the following statistical experiment is performed: somebody flips a coin, repeatedly, until head comes up. This process is described by the recursion equation $X = flip.(tail.X + head)$. Suppose further that the experiment takes place in a closed room, and all information to be obtained about the process in the room is that we can hear the experimenter shout joyfully: 'Head!'. That is, we observe the process $\tau_I(X)$ where $I = \{flip, tail\}$. Now, if the coin is 'fair', it is to be expected that sooner or later (i.e., after a τ-step) the action 'head' will be perceived. Hence, intuitively, $\tau_I(X) = \tau.head$. (This vivid example is from VAANDRAGER [33].)

Koomen's Fair Abstraction Rule (KFAR) is an algebraic rule enabling us to arrive at such a conclusion formally. (For an extensive analysis of this rule see [5].) The simplest form is

$$\frac{x = ix+y \quad (i \in I)}{\tau_I(x) = \tau.\tau_I(y)} \quad \text{KFAR}_1$$

So, KFAR$_1$ expresses the fact that the 'τ-loop' (originating from the i-loop) in $\tau_I(x)$ will not be taken infinitely often. In case this 'τ-loop' is of length 2, the same conclusion is expressed in the rule

$$\frac{x_1 = i_1 x_2 + y_1, x_2 = i_2 x_1 + y_2 \quad (i_1, i_2 \in I)}{\tau_I(x_1) = \tau.\tau_I(y_1 + y_2)} \quad \text{KFAR}_2$$

and it is not hard to guess what the general formulation (KFAR$_n$, $n \geq 1$) will be (see Table 22 in Section 11). In fact, as observed by VAANDRAGER in [33], KFAR$_n$ can already be derived from KFAR$_1$ (at least in the framework of ACP$_\tau^\#$, to be discussed below).

KFAR is of great help in protocol verifications. An example is given in Section 14, where KFAR is used to abstract from a cycle of internal steps which is due to a defective communication channel; the underlying fairness assumption is that this channel is not defective forever, but will function properly after an undetermined period of time. (Just as in the coin flipping experiment the wrong option, tail, is not chosen infinitely often.)

An interesting peculiarity of the present framework is the following. Call the process $\tau^\omega (= \tau.\tau.\tau.....)$ *livelock*. Formally, this is the process $\tau_{\{i\}}(x)$ where x is uniquely defined by the recursion equation $X = i.X$. Noting that $x = i.x = i.x + \delta$ and applying KFAR$_1$ we obtain $\tau^\omega = \tau_{\{i\}}(x) = \tau\delta$. In words: *livelock = deadlock*. There are other semantical frameworks for processes, also in the scope of process algebra but not in the scope of this paper, where this equality does not hold (see [17]).

11. ACP$_\tau^\#$, A FRAMEWORK FOR PROCESS SPECIFICATION AND VERIFICATION

We have now arrived at a framework which will be called ACP$_\tau^\#$, and which contains all the axioms and proof rules introduced so far. In Table 20 the list of all components of ACP$_\tau^\#$ is given; Table 21 contains the equational system ACP$_\tau$ and Table 22 contains the extra features leading first to, as we will call it, ACP$_\tau^+$ and furthermore containing the proof principles which were just introduced, leading to ACP$_\tau^\#$. Note that for *specification* purposes one only needs ACP$_\tau$ or ACP$_\tau^+$; for *verification* one will need ACP$_\tau^\#$ (an extensive example is given in Section 12). Also, it is important to notice that this framework resides entirely on the level of syntax and formal specifications and verification using that syntax - even though some proof rules are infinitary. No semantics for ACP$_\tau^\#$ has been provided yet; this will be done in Section 13. The idea is that 'users' can stay in the realm of this formal system and execute algebraical manipulations, without the need for an excursion into the semantics. That this can be done is demonstrated by the verification of a simple protocol in the next section; at that point the semantics of ACP$_\tau^\#$ (in the form of some model) has, on purpose, not yet been provided. This does not mean that the semantics is unimportant; it does mean that the user needs only be concerned with formula manipulation. The underlying semantics is of great interest for the theory, if only to guarantee the consistency of the formal system; but applications should not be burdened with it, in our intention.

$ACP_\tau^\#$	
BASIC PROCESS ALGEBRA	A1-5
DEADLOCK	A6,7
COMMUNICATION FUNCTION	C1-3
MERGE WITH COMMUNICATION	CM1-9
ENCAPSULATION	D1-4
SILENT STEP	T1-3
SILENT STEP: AUXILIARY AXIOMS	TM1,2; TC1-4
ABSTRACTION	DT; TI1-5
RENAMING	RN
PROJECTION	PR1-4
HAND SHAKING	HA
STANDARD CONCURRENCY	SC
EXPANSION THEOREM	ET
ALPHABET CALCULUS	CA
RECURSIVE DEFINITION PRINCIPLE	RDP
RECURSIVE SPECIFICATION PRINCIPLE	RSP
WEAK APPROXIMATION INDUCTION PRINCIPLE	AIP$^-$
KOOMEN'S FAIR ABSTRACTION RULE	KFAR

TABLE 20

The system up to the first double bar is ACP; up to the second double bar we have ACP_τ, and up to the third double bar, ACP_τ^+.

		ACP_τ		
$x+y=y+x$	A1	$x\tau=x$		T1
$x+(y+z)=(x+y)+z$	A2	$\tau x+x=\tau x$		T2
$x+x=x$	A3	$a(\tau x+y)=a(\tau x+y)+ax$		T3
$(x+y)z=xz+yz$	A4			
$(xy)z=x(yz)$	A5			
$x+\delta=x$	A6			
$\delta x=\delta$	A7			
$a\vert b=b\vert a$	C1			
$(a\vert b)\vert c=a\vert(b\vert c)$	C2			
$\delta\vert a=\delta$	C3			
$x\Vert y=x\mathbin{\Vert\mkern-2mu\relbar} y+y\mathbin{\Vert\mkern-2mu\relbar} x+x\vert y$	CM1			
$a\mathbin{\Vert\mkern-2mu\relbar} x=ax$	CM2	$\tau\mathbin{\Vert\mkern-2mu\relbar} x=\tau x$		TM1
$ax\mathbin{\Vert\mkern-2mu\relbar} y=a(x\Vert y)$	CM3	$\tau x\mathbin{\Vert\mkern-2mu\relbar} y=\tau(x\Vert y)$		TM2
$(x+y)\mathbin{\Vert\mkern-2mu\relbar} z=x\mathbin{\Vert\mkern-2mu\relbar} z+y\mathbin{\Vert\mkern-2mu\relbar} z$	CM4	$\tau\vert x=\delta$		TC1
$ax\vert b=(a\vert b)x$	CM5	$x\vert\tau=\delta$		TC2
$a\vert bx=(a\vert b)x$	CM6	$\tau x\vert y=x\vert y$		TC3
$ax\vert by=(a\vert b)(x\Vert y)$	CM7	$x\vert\tau y=x\vert y$		TC4
$(x+y)\vert z=x\vert z+y\vert z$	CM8			
$x\vert(y+z)=x\vert y+x\vert z$	CM9	$\partial_H(\tau)=\tau$		DT
		$\tau_I(\tau)=\tau$		TI1
$\partial_H(a)=a$ if $a\notin H$	D1	$\tau_I(a)=a$ if $a\notin I$		TI2
$\partial_H(a)=\delta$ if $a\in H$	D2	$\tau_I(a)=\tau$ if $a\in I$		TI3
$\partial_H(x+y)=\partial_H(x)+\partial_H(y)$	D3	$\tau_I(x+y)=\tau_I(x)+\tau_I(y)$		TI4
$\partial_H(xy)=\partial_H(x)\cdot\partial_H(y)$	D4	$\tau_I(xy)=\tau(x)\cdot\tau_I(y)$		TI5

TABLE 21

TABLE 22

REMAINING AXIOMS AND RULES FOR $ACP_\tau^\#$

$\rho_f(a)=f(a)$	RN1	$\pi_1(ax)=a$	PR1
$\rho_f(x+y)=\rho_f(x)+\rho_f(y)$	RN2	$\pi_{n+1}(ax)=a.\pi_n(x)$	PR2
$\rho_f(xy)=\rho_f(x).\rho_f(y)$	RN3	$\pi_n(a)=a$	PR3
$\rho_{id}(x)=x$	RN4	$\pi_n(x+y)=\pi_n(x)+\pi_n(y)$	PR4
$(\rho_f\circ\rho_g)(x)=\rho_{f\circ g}(x)$	RN5	$\pi_n(\tau)=\tau$	PR5
$\rho_f(\tau)=\tau$	RN6	$\pi_n(\tau x)=\tau.\pi_n(x)$	PR6

$x\mid y\mid z=\delta$	HA
$x\mid y=y\mid x$	SC1
$x\parallel y=y\parallel x$	SC2
$x\mid(y\mid z)=(x\mid y)\mid z$	SC3
$(x\mathbin{\lfloor\!\lfloor} y)\mathbin{\lfloor\!\lfloor} z=x\mathbin{\lfloor\!\lfloor}(y\parallel z)$	SC4
$(x\mid ay)\mathbin{\lfloor\!\lfloor} z=x\mid(ay\mathbin{\lfloor\!\lfloor} z)$	SC5
$x\parallel(y\parallel z)=(x\parallel y)\parallel z$	SC6

$$x_1\parallel\ldots\parallel x_n = \sum_{1\leq i\leq n} x_i\mathbin{\lfloor\!\lfloor}(\mathop{\parallel}_{\substack{1\leq k\leq n\\ k\neq i}} x_k) \qquad \text{ET}$$

$$+ \sum_{1\leq i<j\leq n}(x_i\mid x_j)\mathbin{\lfloor\!\lfloor}(\mathop{\parallel}_{\substack{1\leq k\leq n\\ k\neq i,k\neq j}} x_k) \quad (n\geq 3)$$

$\alpha(\delta)=\varnothing$	AB1
$\alpha(\tau)=\varnothing$	AB2
$\alpha(a)=\{a\}$ (if $a\neq\delta$)	AB3
$\alpha(\tau x)=\alpha(x)$	AB4
$\alpha(ax)=\{a\}\cup\alpha(x)$ (if $a\neq\delta$)	AB5
$\alpha(x+y)=\alpha(x)\cup\alpha(y)$	AB6
$\alpha(x)=\cup_{n\geq 1}\alpha(\pi_n(x))$	AB7
$\alpha(\tau_I(x))=\alpha(x)-I$	AB8

$\alpha(x)\mid(\alpha(y)\cap H)\subseteq H\Rightarrow\partial_H(x\parallel y)=\partial_H(x\parallel\partial_H(y))$	CA1
$\alpha(x)\mid(\alpha(y)\cap I)=\varnothing\Rightarrow\tau_I(x\parallel y)=\tau_I(x\parallel\tau_I(y))$	CA2
$\alpha(x)\cap H=\varnothing\Rightarrow\partial_H(x)=x$	CA3
$\alpha(x)\cap I=\varnothing\Rightarrow\tau_I(x)=x$	CA4

RDP	Every guarded and abstraction free specification has a solution
RSP	Every guarded and abstraction free specification has at most one solution
AIP$^-$	Every process which has an abstraction free specification is determined by its finite projections

$$\frac{\forall n\in\mathbb{Z}_k \quad x_n=i_n.x_{n+1}+y_n \quad (i_n\in I)}{\tau_I(x_n)=\tau.\tau_I(\Sigma_{m\in\mathbb{Z}_k}y_m)} \quad \text{KFAR}_k$$

It should be noted that there is redundancy in this presentation; as we already stated, AIP$^-$ implies RSP and there are other instances where we can save some axioms or rules (for instance, the projection axioms PR1-6 turn out to be definable from the other operators). This would however not enhance clarity. Also note that one of the standard concurrency axioms, SC5, is different (namely more restrictive) than the corresponding one for the situation without τ in Table 9 (the second axiom).

So ACP$_\tau^\#$ is a medium for formal process specifications and verifications; let us note that we also admit infinite specifications. As the system is meant to have practical applications, we will only encounter *computable* specifications. A finite specification (of which an expression is a particular case) is trivially computable; an infinite specification $\{E_n | n \geq 0\}$ where E_n is the recursion equation $X_n = T(X_1, \ldots, X_{f(n)})$ is computable if after some coding, in which E_n is coded as a natural number e_n, the sequence $\{e_n | n \geq 0\}$ is computable. Here an important question arises: *is every computable specification provably equal to a finite specification?* At present we are unable to answer this question; but we can state that the answer is affirmative *relative to certain models* of ACP$_\tau^\#$. Before we elaborate this, a verification of a simple protocol is demonstrated.

12. AN ALGEBRAIC VERIFICATION OF THE ALTERNATING BIT PROTOCOL

In this section we will demonstrate a verification of a simple communication protocol, the Alternating Bit Protocol, in the framework of ACP$_\tau^\#$. (In fact, not all of ACP$_\tau^\#$ is needed.) This verification is from [13]; the present streamlined treatment was kindly made available to us by F.W. VAANDRAGER (CWI Amsterdam).

Let D be a finite set of data. Elements of D are to be transmitted by the ABP from port 1 to port 2. The ABP can be visualized as follows:

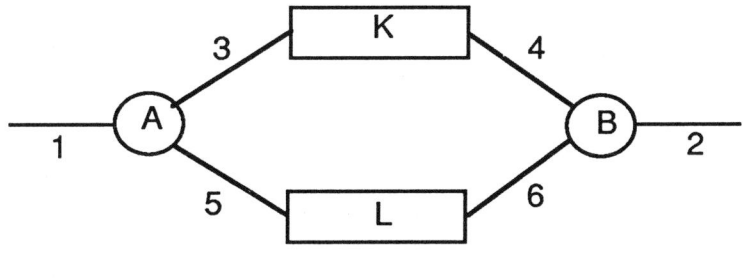

FIGURE 3

There are four components:
A: Reads a Message (RM) at 1. Thereafter it Sends a Frame (SF), consisting of the message and a control bit, into channel K until a correct

Acknowledgement has been Received (RA) via channel L. The equations for A are as follows. We will always use the notations: datum $d\in D$, bit $b\in\{0,1\}$, frame $f\in D\times\{0,1\}$ (so a frame f is of the form db).

$$A = RM^0$$
$$RM^b = \Sigma_d\, r1(d).SF^{db}$$
$$SF^{db} = s3(db).RA^{db}$$
$$RA^{db} = (r5(1-b) + r5(e)).SF^{db} + r5(b).RM^{1-b}$$

K: data transmission channel K communicates elements of $D\times\{0,1\}$, and may communicate these correctly or communicate an error value e. K is supposed to be fair in the sense that it will not produce an infinite consecutive sequence of error outputs.

$$K = \Sigma_f\, r3(f).K^f$$
$$K^f = (\tau.s4(e) + \tau.s4(f)).K$$

The τ's in the second equation express that the choice whether or not a frame f is to be communicated correctly, cannot be influenced by one of the other components.

B: Receives a Frame (RF) via channel K. If the control bit of the frame is OK, then the Message is Sent (SM) at 2. B Sends back Acknowledgement (SA) via L.

$$B = RF^0$$
$$RF^b = (\Sigma_d\, r4(d(1-b)) + r4(e)).SA^{1-b} + \Sigma_d\, r4(db).SM^{db}$$
$$SA^b = s6(b).RF^{1-b}$$
$$SM^{db} = s2(d).SA^b$$

L: the task of acknowledgement transmission channel L is to communicate boolean values from B to A. The channel L may yield error outputs but is also supposed to be fair.

$$L = \Sigma_b\, r6(b).L^b$$
$$L^b = (\tau.s5(e) + \tau.s5(b)).L$$

Define $\mathbf{D} = D \cup (D\times\{0,1\}) \cup \{0,1\} \cup \{e\}$. $\mathbf{D}$ is the set of 'generalized' data (i.e. plain data, frames, bits, error) that occur as parameter of atomic actions. We use the notation: $g\in\mathbf{D}$. For $t\in\{1,2,\ldots,6\}$ there are send, read, and communication actions:

$$A = \{st(g), rt(g), ct(g) | g\in\mathbf{D}, t\in\{1,2,\ldots,6\}\}.$$

We define communication by $st(g)|rt(g) = ct(g)$ for $g\in\mathbf{D}$, $t\in\{1,2,\ldots,6\}$ and all other communications give δ. Define the following two subsets of A:

$$H = \{st(g), rt(g) | t\in\{3,4,5,6\}, g\in\mathbf{D}\}$$
$$I = \{ct(g) | t\in\{3,4,5,6\}, g\in\mathbf{D}\}.$$

Now the ABP is described by ABP $= \tau_I \circ \partial_H(A\|K\|B\|L)$. The fact that this is a correct protocol is asserted by

THEOREM 7. $ACP_\tau^\# \vdash ABP = \Sigma_d\, r1(d).s2(d).ABP$.

(Actually, we need only the part of $ACP_\tau^\#$ consisting of $ACP_\tau + SC + RDP + RSP + CA + KFAR$ - see Tables 21, 22.)

PROOF. Let $I' = \{ct(g) | t \in \{3,4,5\}, f \in \mathbf{D}\}$. We will use $[x]$ as a notation for $\tau_{I'} \circ \partial_H(x)$. Consider the following system of recursion equations:

$$
\begin{array}{l}
(0)\ X = X_1^0 \\
(1)\ X_1^b = \Sigma_d\, r1(d).X_2^{db} \\
(2)\ X_2^{db} = \tau.X_3^{db} + \tau.X_4^{db} \\
(3)\ X_3^{db} = c6(1-b).X_2^{db} \\
(4)\ X_4^{db} = s2(d).X_5^{db} \\
(5)\ X_5^{db} = c6(b).X_6^{db} \\
(6)\ X_6^{db} = \tau.X_5^{db} + \tau.X_1^{1-b}
\end{array}
$$

We claim that $ACP_\tau^\# \vdash X = [A \| K \| B \| L]$. We prove this by showing that $[A \| K \| B \| L]$ satisfies the same recursion equations (0)-(6) as X does. In the computations below, the bold-face part denotes the part of the expression currently being 'rewritten'.

$[A \| K \| B \| L] = [RM^0 \| K \| RF^0 \| L]$ \hfill (0)

$[\mathbf{RM^b} \| K \| RF^b \| L] = \Sigma_d\, r1(d).[\mathbf{SF^{db}} \| \mathbf{K} \| RF^b \| L]$ \hfill (1)

$\qquad = \Sigma_d\, r1(d).\tau.[RA^{db} \| K^{db} \| RF^b \| L]$

$\qquad = \Sigma_{d \in D}\, r1(d).[RA^{db} \| K^{db} \| RF^b \| L]$

$[RA^{db} \| \mathbf{K^{db}} \| RF^b \| L] = \tau.[RA^{db} \| \mathbf{s4(e).K} \| \mathbf{RF^b} \| L]$ \hfill (2)

$\qquad + \tau.[RA^{db} \| \mathbf{s4(db).K} \| \mathbf{RF^b} \| L] = \tau.[RA^{db} \| K \| SA^{1-b} \| L]$

$\qquad + \tau.[RA^{db} \| K \| SM^{db} \| L]$

$[RA^{db} \| K \| \mathbf{SA^{1-b}} \| \mathbf{L}] = c6(1-b).[RA^{db} \| K \| RF^b \| \mathbf{L^{1-b}}]$ \hfill (3)

$\qquad = c6(1-b).(\tau.[\mathbf{RA^{db}} \| K \| RF^b \| \mathbf{s5(e).L}]$

$\qquad + \tau.[\mathbf{RA^{db}} \| K \| RF^b \| \mathbf{s5(1-b).L}])$

$\qquad = c6(1-b).\tau.[\mathbf{SF^{db}} \| \mathbf{K} \| RF^b \| L]$

$\qquad = c6(1-b).\tau.\tau.[RA^{db} \| K^{db} \| RF^b \| L]$

$\qquad = c6(1-b).[RA^{db} \| K^{db} \| RF^b \| L].$

$[RA^{db} \| K \| \mathbf{SM^{db}} \| L] = s2(d).[RA^{db} \| K \| SA^b \| L].$ \hfill (4)

$[RA^{db} \| K \| \mathbf{SA^b} \| L] = c6(b).[RA^{db} \| K \| RF^{1-b} \| L^b].$ \hfill (5)

$[RA^{db} \| K \| RF^{1-b} \| \mathbf{L^b}] = \tau.[\mathbf{RA^{db}} \| K \| RF^{1-b} \| \mathbf{s5(e).L}]$ \hfill (6)

$$+ \tau.[RA^{db} \| K \| RF^{1-b} \| s5(b).L]$$
$$= \tau.[SF^{db} \| K \| RF^{1-b} \| L]$$
$$+ \tau.[RM^{1-b} \| K \| RF^{1-b} \| L].$$
$$[SF^{db} \| K \| RF^{1-b} \| L] = \tau.[RA^{db} \| K^{db} \| RF^{1-b} \| L] \qquad (7)$$
$$= \tau.(\tau[RA^{db} \| s4(e).K \| RF^{1-b} \| L]$$
$$+ \tau.[RA^{db} \| s4(db).K \| RF^{1-b} \| L])$$
$$= \tau.[RA^{db} \| K \| SA^{b} \| L].$$

Now substitute (7) in (6) and apply RSP + RDP. Using the conditional axioms (see Table 22, Section 11) we have ABP $= \tau_I(X) = \tau_I(X_1^0)$. Further, an application of KFAR$_2$ gives $\tau_I(X_2^{db}) = \tau.\tau_I(X_4^{db})$ and $\tau_I(X_5^{db}) = \tau.\tau_I(X_1^{1-b})$. Hence,

$$\tau_I(X_1^b) = \Sigma_d \, r\,1(d).\tau_I(X_2^{db}) = \Sigma_d \, r\,1(d).\tau_I(X_4^{db})$$
$$= \Sigma_d \, r\,1(d).s\,2(d).\tau_I(X_5^{db}) = \Sigma_d \, r\,1(d).s\,2(d).\tau_I(X_1^{1-b})$$

and thus

$$\tau_I(X_1^0) = \Sigma_d \, r\,1(d).s\,2(d).\Sigma_{d'} r\,1(d').s\,2(d').\tau_I(X_1^0)$$
$$\tau_I(X_1^1) = \Sigma_d \, r\,1(d).s\,2(d).\Sigma_{d'} r\,1(d').s\,2(d').\tau_I(X_1^1).$$

Applying RDP + RSP gives $\tau_I(X_1^0) = \tau_I(X_1^1)$ and therefore $\tau_I(X_1^0) = \Sigma_d r\,1(d).s\,2(d).\tau_I(X_1^0)$, which finishes the proof of the theorem. □

More complicated communication protocols have been verified in ACP$_\tau^\#$ recently by VAANDRAGER [33]: a Positive Acknowledgement with Retransmission protocol and a One Bit Sliding Window protocol. There the notion of *redundancy in a context* is used as a tool which facilitates the verifications. A related method, using a modular approach, is employed in KOYMANS and MULDER [26], where a version of the Alternating Bit Protocol called the Concurrent Alternating Bit Protocol is verified in ACP$_\tau^\#$. (In fact, also in the verifications in [26], [33] one only needs the part of ACP$_\tau^\#$ mentioned after Theorem 7.)

13. THE GRAPH MODEL FOR ACP$_\tau^\#$

We will give a quick introduction to what we consider to be the 'main' model of ACP$_\tau^\#$. The basic building material consists of the domain of *countably branching, labeled, rooted, connected, directed multigraphs*. Such a graph, also called a process graph, consists of a possibly infinite set of nodes s with one distinguished node s_0, the root. The edges, also called transitions or steps, between the nodes are labeled with an element from the action alphabet; also δ and τ may be edge labels. We use the notation $s \to_a t$ for an a-transition from node s to node t; likewise $s \to_\tau t$ is a τ-transition and $s \to_\delta t$ is a δ-step. That the graph is connected means that every node must be accessible by finitely many steps from the root node.

Corresponding to the operations $+, ., \|, \mathbin{\|\mkern-6mu\raise1pt\hbox{$\llcorner$}}, |, \partial_H, \tau_I, \pi_n, \alpha$ in $\text{ACP}_\tau^{\#}$ we define operations in this domain of process graphs. Precise definitions can be found in [1,5]; we will sketch some of them here. The sum $g+h$ of two process graphs g,h is obtained by glueing together the roots of g and h (see Figure 4(i)); there is one caveat: if a root is cyclic (i.e. lying on a cycle of transitions leading back to the root), then the initial part of the graph has to be 'unwound' first so as to make the root acyclic (see Figure 4(ii)). The product $g.h$ is obtained by appending copies of h to each terminal node of g; alternatively, one may first identify all terminal nodes of g and then append one copy of h to the unique terminal node if it exists (see Figure 4 (iii)). The merge $g\|h$ is obtained as a cartesian product of both graphs, with 'diagonal' edges for communications (see Figure 4(v) for an example without communication, and Figure 4(vi) for an example with communication action $a|b$. Definitions of the auxiliary operators are somewhat more complicated and not discussed here. The encapsulation and abstraction operators are simply renamings, that replace the edge labels in H resp. in I by δ resp. τ. Definitions of the projection operators π_n and α should be clear from the axioms by which they are specified. As to the projection operators, it should be emphasized that τ-steps are 'transparent': they do not increase the depth.

FIGURE 4

OPERATIONS ON PROCESS GRAPHS

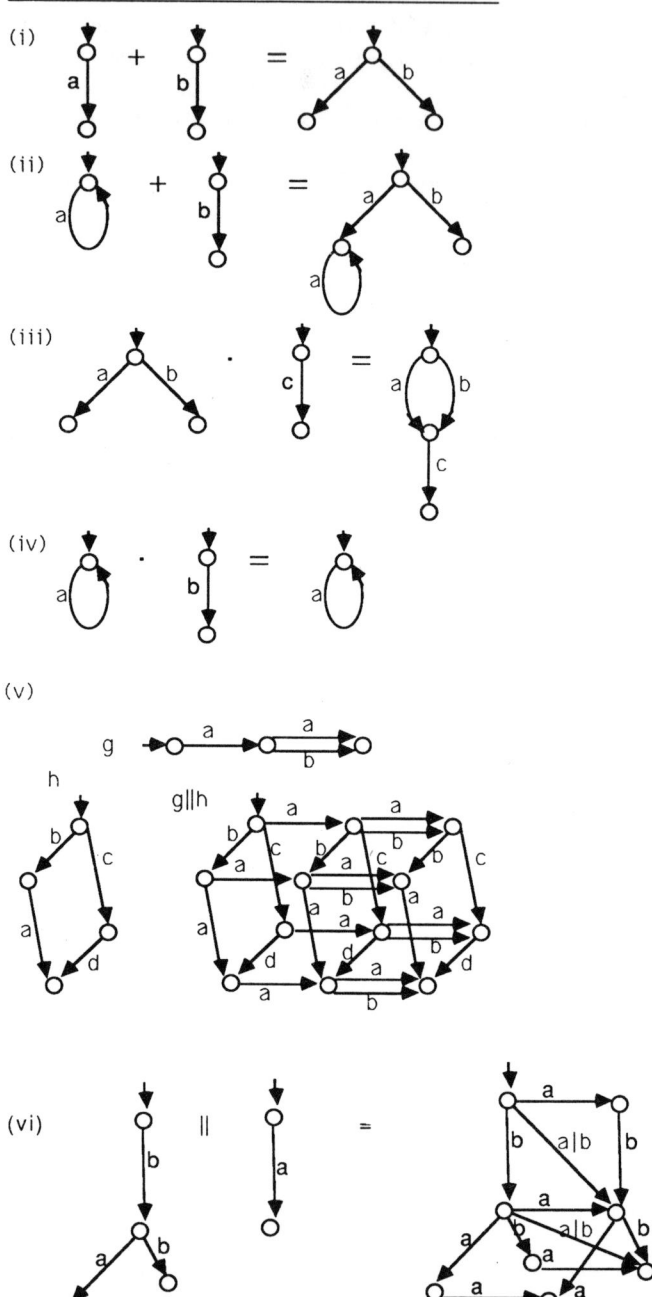

This domain of process graphs equipped with the operations just introduced, is not yet a model of ACP_τ: for instance the axiom $x+x=x$ does not hold. In order to obtain a model, we define an equivalence on the process graphs which is moreover a congruence w.r.t. the operations. This equivalence is called *bisimulation congruence* or *bisimilarity*. (The original notion is due to PARK [32]; it was anticipated by Milner's observational equivalence, see [30].) In order to define this notion, let us first introduce the notation $s \Rightarrow_a t$ for nodes s, t of graph g, indicating that from node s to node t there is a finite path consisting of zero or more τ-steps and one a-step followed by zero or more τ-steps. Let us say that in this situation there is a 'generalized a-step' from s to t. Likewise with 'a' replaced by 'τ'. Next, let a *coloring* of process graph g be a surjective mapping from a set of 'colors' C to the node set of g, such that the color assigned to the root of g is different from all other colors, and furthermore, such that all end nodes are assigned the same color which is different from other colors. Now two process graphs g, h are bisimilar if there are colorings of g, h such that (1) the roots of g, h have the same color and (2) whenever *somewhere* in the two graphs a generalized a-step is possible from a node with color c to a node with color c', then *every* c-colored node admits a generalized a-step to a c'-colored node (be it in g or in h). We use the notation $g \underline{\leftrightarrow} h$ to indicate that g, h are bisimilar. One can prove that $\underline{\leftrightarrow}$ is a congruence and, if **G** is the original domain of countably branching process graphs:

THEOREM 8 ([5]). $\mathbf{G}/\underline{\leftrightarrow}$ *is a model of* $ACP_\tau^\#$.

Remarkably, this graph model (as we will call it henceforth) does not satisfy the unrestricted Approximation Induction Principle. A counterexample is given (in a self-explaining notation) by the two graphs $g = \Sigma_{n \geq 1} a^n$ and $h = \Sigma_{n \geq 1} a^n + a^\omega$; while g and h have the same finite projections $\pi^n(g) = \pi^n(h) = a + a^2 + a^3 + \ldots + a^n$, they are not bisimilar due to the presence of the infinite trace of a-steps in h. It might be thought that it would be helpful to restrict the domain of process graphs to finitely branching graphs, in order to obtain a model which does satisfy AIP, but there are two reasons why this is not the case: (1) the finitely branching graph domain would not be closed under the operations, in particular the communication merge ($|$); (2) a similar counterexample can be obtained by considering the finitely branching graphs $g' = \tau_{\{t\}}(g'')$ where g'' is the graph defined by $\{X_n = a^n + tX_{n+1} | n \geq 1\}$ and $h' = g' + a^\omega$.

14. THE EXPRESSIVE POWER OF ACP_τ

ACP_τ is a powerful specification mechanism; in a sense it is a universal specification mechanism: *every finitely branching, computable process can be finitely specified* in ACP_τ. We have to be more precise about the notion of 'computable process'. First, an intuitive explanation: suppose a finitely branching process graph g is actually given; the labels may include τ, and there may be even infinite τ-traces. That g is 'actually' given means that the process graph g must be 'computable': a finite recipe describes the graph, in the form of a coding of the nodes in natural numbers and recursive functions giving in-degree, out-degree, edge-labels. This notion of a computable process graph is rather obvious, and we will not give details of the definition here (these can be found in [5]).

Now even if g is an infinite process graph, it can be specified by an infinite computable specification, as follows. First rename all τ-edges in g to t-edges, for a 'fresh' atom t. Call the resulting process graph: g_t. Next assign to each node s of g_t a recursion variable X_s and write down the recursion equation for X_s according to the outgoing edges of node s. Let X_{s0} be the variable corresponding to the root s_0 of g_t. As g is computable, g_t is computable and the resulting 'direct' specification $E = \{X_s = T_s(\mathbf{X}) | s \in NODES(g_t)\}$ is evidently also computable (i.e.: the nodes can be numbered as $s_n (n \geq 0)$ and after coding the sequence e_n of codes of equations $E_n : X_{sn} = T_{sn}(\mathbf{X})$ is a computable sequence). Now the specification which uniquely determines g is simply: $\{Y = \tau_{\{t\}}(X_{s0})\} \cup E$. In fact all specifications below will have the form $\{X = \tau_I(X_0), X_n = T_n(\mathbf{X}) | n \geq 0\}$ where the guarded expressions $T_n(\mathbf{X})(= T_n(X_{i1}, \ldots, X_{in}))$ contain no abstraction operators τ_J. They may contain all other process operators. We will say that such specifications have *restricted abstraction*.

However, we want more than a computable specification with restricted abstraction: to describe process graph g we would like to find a *finite* specification with restricted abstraction for g. Indeed this is possible:

THEOREM 9. *Let the finitely branching and computable process graph g determine $\tilde{g}$ in the graph model of ACP_τ. Then there is a finite specification with restricted abstraction E in ACP_τ such that $[\![E]\!] = \tilde{g}$.*

Here $[\![E]\!]$ is the semantics of E in the graph model. The proof in [5] is by constructing a Turing machine in ACP_τ; the 'tape' is obtained by glueing together two stacks. A stack has a simple finite specification, already in BPA; see [15]. A stronger fact would be the assertion that every computable specification with restricted abstraction in ACP_τ is provably equivalent (in $ACP_\tau^{\#}$) to a finite specification with restricted abstraction. At present we do not know whether this is true.

It should be noted that abstraction plays an essential role in this finite specification theorem. If $f : \mathbb{N} \to \{a, b\}$ is a sequence of a, b, let p_f be the process $f(0).f(1).f(2)\ldots$ (more precisely: the unique solution of the specification $\{X_n = f(n).X_{n+1} | n \geq 0\}$). Now:

THEOREM 10. *There is a computable function f such that process p_f is not definable by a finite specification without abstraction operator.*

A fortiori, p_f is not finitely definable in ACP. The proof in [5] is via a simple diagonalization argument.

The finite specification theorem, which is relative to the graph model of $ACP_\tau^{\#}$, in fact generalizes to the class of 'extensional' models. In order to define this concept we first define the notion of 'canonical process graph' of a process in an arbitrary process algebra.

Let $\mathcal{C}$ be a process algebra (i.e. a model of the axiom system under consideration, in casu ACP_τ). Let $p,q \in \mathcal{C}$. We define *transition relations* $\to_a$, for every atomic action a, and $\to_\tau$, as follows: $p\to_a q$ iff $p=a.q+r$ for some r. Moreover, if $p=a+r$ for some r, then $p\to_a o$ where o is an auxiliary element not in the domain of $\mathcal{C}$. The same with τ instead of 'a'. Now the *canonical process graph* of p (notation: $G(p)$) is the labeled and directed graph with root: p and with nodes all elements accessible from p via the transitions $\to_a, \to_\tau$. The edges of the canonical process graph are given by the transitions. Note that every element in every process algebra thus has a canonical process graph. In analogy with the situation in set theory, we will call a process algebra *extensional* if whenever p,q have the same process graph, they are equal. (Cf. the 'observable' process spaces in HESSELINK [22].) In an extensional model an element is fully determined by its transition relations to other elements. The models that we have introduced are all extensional. A process is *finitely branching* when its canonical graph is. Now we can define that a process is *computable* when its canonical graph is. The finite specification theorem above generalizes to:

THEOREM 11. *Let p be a finitely branching, computable process in an extensional process algebra (a model of ACP_τ). Then p can, in ACP_τ, be specified by a finite specification with restricted abstraction.*

It should be possible to remove the assumption 'finitely branching' in favour of 'countably branching', but we will not attempt to do so here.

15. A FUNDAMENTAL INCOMPATIBILITY

As we have seen, the graph model of $ACP_\tau^\#$ (Section 13) does not satisfy the unrestricted Approximation Induction Principle which states that every process is uniquely determined by its finite projections. It is natural to search for a model in which this principle does hold. However, R.J. VAN GLABBEEK (CWI Amsterdam) recently noticed that such a model does not exist, if one wishes to adhere to the very natural assumption that composition of abstraction operators is commutative. As always, we refer here to models which are trace consistent. We will consider the following consequence of RN5 in Table 22: $\tau_{\{a\}} \circ \tau_{\{b\}} = \tau_{\{b\}} \circ \tau_{\{a\}}$ which we will denote by CA (commutativity of abstraction). Now we have:

THEOREM 12 ([21]). $ACP_\tau + KFAR_1 + RDP + RSP + CA + AIP \vdash \tau = \tau + \tau\delta$.

By way of exception we include the interesting proof. Consider the specifications $E = \{X_n = aX_{n+1} + b^n | n \geq 0\}$, $F = \{Y = bY\}$ and $G = \{Z = aZ + \tau\}$. By RDP+RSP we have unique solutions for these specifications; they will be denoted by $X_n (n \geq 0)$, Y, Z. By $KFAR_1$ we have immediately: $\tau_{\{b\}}(\underline{Y}) = \tau\delta$ and $\tau_{\{a\}}(\underline{Z}) = \tau.\overline{\tau} = \overline{\tau}.$ Further, $\tau_{\{b\}}(\underline{X_n}) =$

$a.\tau_{\{b\}}(X_{n+1})+\tau^n = a.\tau_{\{b\}}(X_{n+1})+\tau$, so the sequence $\{\tau_{\{b\}}(X_n)|n \geq 0\}$ is a solution of the infinite specification $G' = \{Z_n = a\underline{Z}_{n+1}+\tau|n \geq 0\}$. Clearly this last specification is also satisfied by the sequence $\{Z, Z, ...\}$. Hence, by RSP, $\tau_{\{b\}}(\underline{X}_n) = \underline{Z}$. It follows that $\tau_{\{b\}}(a\underline{X}_0) = a\underline{Z}$; whence

$$\tau_{\{a\}} \circ \tau_{\{b\}}(a\underline{X}_0) = \tau_{\{a\}}(a\underline{Z}) = \tau.\tau_{\{a\}}(\underline{Z}) = \tau.\tau = \tau. \qquad (1)$$

Now, using the τ-law T2 and in particular its consequence $\tau(x+y) = \tau(x+y)+x$, one proves easily that for all k:

$$\tau_{\{a\}}(a\underline{X}_1) = \tau_{\{a\}}(a\underline{X}_1) + b^k.$$

(E.g. for $k=0$: $\tau_{\{a\}}(a\underline{X}_0) = \tau.\tau_{\{a\}}(\underline{X}_0) = \tau.\tau_{\{a\}}(a\underline{X}_1 + b) = \tau(\tau_{\{a\}}(a\underline{X}_1) + b) = \tau(\tau_{\{a\}}(a\underline{X}_1) + b) + b = \tau_{\{a\}}(a\underline{X}_0) + b).$

So $\pi_k(\tau_{\{a\}}(a\underline{X}_0)) = \pi_k(\tau_{\{a\}}(a\underline{X}_0)) + b^k = \pi_k(\tau_{\{a\}}(a\underline{X}_0) + Y)$, for all k. Therefore by AIP: $\tau_{\{a\}}(a\underline{X}_0) = \tau_{\{a\}}(a\underline{X}_0) + Y$. Hence, using (1) and CA:

$$\tau_{\{b\}} \circ \tau_{\{a\}}(a\underline{X}_0) = \tau_{\{b\}} \circ \tau_{\{a\}}(a\underline{X}_0) + \tau_{\{b\}}(Y) = \tau_{\{a\}} \circ \tau_{\{b\}}(a\underline{X}_0) + \tau_{\{b\}}(Y)$$
$$= \tau + \tau\delta, \qquad (2)$$

and again by (1) and CA: $\tau = \tau + \tau\delta$. □

So, in every theory extending ACP_τ, the combination of features AIP, KFAR, CA, RDP+RSP is impossible. Among such theories are also theories where the equivalence on processes is much coarser, such as in Hoare's well-known failure model [25]; this semantics is not discussed in the present paper. VAN GLABBEEK [21] moreover notices that there is quite a subtle trade-off between these four features. In the graph model of $ACP_\tau^\#$ we have AIP^-, KFAR, CA, RDP+RSP. There is also a failure model satisfying AIP, $KFAR^-$, CA, RDP+RSP, where $KFAR^-$ is a restricted form of KFAR (see [17]). In fact it seems that models can be found by weakening any of the four features that make up the impossible combination.

16. Additional features

As we have seen in Section 14, ACP_τ is a universal specification system for (finitely branching) computable processes. Yet this does not preclude the search for additional operators on processes, in order to make finite specifications of computable processes not only theoretically possible, but also practically feasible. The two main additional operators which have been defined and studied in process algebra are the *priority operator* and the *state operator*.

By means of the priority operator θ one can enforce that certain actions are privileged and have priority over others. Thus θ is parameterized by a partial order $>$ on the set of atomic actions; the constant δ (deadlock) will always be the least element in this partial ordering. As an example, let atomic actions a, b, c be ordered by: $a, b < c$. Then $\theta(a+b+c) = c$, $\theta(a+b) = a+b$, $\theta(ax+cy) = c\theta(y)$. Using an auxiliary operator $\triangleleft$ ('unless') we can axiomatize θ in finitely many equations:

PRIORITY OPERATOR
$a \triangleleft b = a$ if $\neg(a < b)$
$a \triangleleft b = \delta$ if $a < b$
$x \triangleleft yz = x \triangleleft y$
$x \triangleleft (y + z) = (x \triangleleft y) \triangleleft z$
$xy \triangleleft z = (x \triangleleft z)y$
$(x + y) \triangleleft z = x \triangleleft z + y \triangleleft z$
$\theta(a) = a$
$\theta(xy) = \theta(x) \cdot \theta(y)$
$\theta(x + y) = \theta(x) \triangleleft y + \theta(y) \triangleleft x$

TABLE 23

The priority operator θ (with its axioms) can be joined with ACP (see Section 11); the result is called ACP_θ. Note that we do not join θ and ACP_τ; at present the interaction between τ and θ is not clear. In [4] an elimination theorem is proved stating that every closed ACP_θ-term is (in ACP_θ) provably equal to a BPA_δ-term, that is a term without occurrences of other operators than $\cdot$ and $+$. Using θ, one can model *interrupts* (see [4]). Another application is given in [9]: there a *put* and *get* mechanism has been modeled using ACP_θ. Communication by means of put and get mechanism differs from the synchronous hand shaking mechanism: even if the 'receiving' process is not enabled to receive the message, the 'sending' process can perform a put action, and proceed with its execution. Likewise, a receiving process can perform a get action even when there is nothing to get, and continue in that case. Using the put mechanism, it is shown in [9] how a *broadcasting* mechanism for arbitrarily many receivers can be modeled.

Another very useful operator is the *state operator* λ_s, where s is some state from a state space S. The essential equation is $\lambda_s(ax) = a' \cdot \lambda_{s'}(x)$. Here s' and a' are the state and action respectively resulting from executing a in state s. The state operator is useful in designing an algebraic semantics for programming languages; when dealing with object-oriented programming languages or specification languages, it is useful to provide the state operator with a name m of the object in question. Thus $\lambda_s^m(x)$ can intuitively be perceived as the process resulting from input x (the 'program') in m (the 'machine') in s (the state of m). Writing $a' = a(m,s)$ (the *action function*) and $s' = s(m,a)$ (the *effect function*) the state operator is axiomatized by:

STATE OPERATOR
$\lambda_s^m(\delta) = \delta$
$\lambda_s^m(a) = a(m,s)$
$\lambda_s^m(ax) = a(m,s) \cdot \lambda_{s(m,a)}^m(x)$
$\lambda_s^m(x+y) = \lambda_s^m(x) + \lambda_s^m(y)$

TABLE 24

In fact, this state operator is a generalization of the renaming operator in Section 6. In [1] asynchronous communication is modeled using the state operator: here a message from sender to receiver may have some delay.

Another mechanism which is of interest for specifications is *process creation*. In [8] axioms for a process creation operator have been given; for some examples of its use see also [1]. A typical example is the modeling of the *sieve of Eratosthenes*.

Finally, we mention the work of VRANCKEN [34] where the *empty process* ϵ has been axiomatized. The basic axioms for this process are $\epsilon x = x$ and $x\epsilon = x$. It should be pointed out that addition of such a process requires careful consideration in order to preserve the consistency of the whole axiomatization. Using this process several short-cuts in process specifications can be obtained.

REFERENCES
1. J.C.M. BAETEN (1986). *Procesalgebra*, course notes (in Dutch), Department of Computer Science, University of Amsterdam.
2. J.C.M. BAETEN, J.A. BERGSTRA (1985). *Global Renaming Operators in Concrete Process Algebra*, CWI Report CS-R8521, Amsterdam.
3. J.C.M. BAETEN, J.A. BERGSTRA, J.W. KLOP (1985). *Conditional Axioms and α/β Calculus in Process Algebra*, CWI Report CS-R8502, Amsterdam.
4. J.C.M. BAETEN, J.A. BERGSTRA, J.W. KLOP (1985). *Syntax and Defining Equations for an Interrupt Mechanism in Process Algebra*, CWI Report CS-R8503, Amsterdam.
5. J.C.M. BAETEN, J.A. BERGSTRA, J.W. KLOP (1985). *On the Consistency of Koomen's Fair Abstraction Rule*, CWI Report CS-R8511, Amsterdam.
6. J.W. DE BAKKER, J.I. ZUCKER (1982). Denotational semantics of concurrency. *Proc. 14th ACM Symp. Theory of Comp.*, 153-158.
7. J.W. DE BAKKER, J.I. ZUCKER (1982). Processes and the denotational semantics of concurrency. *Information and Control 54 (1/2)*, 70-120.
8. J.A. BERGSTRA (1985). *A Process Creation Mechanism in Process Algebra*, Logic Group Preprint Series Nr. 2, Dept. of Philosophy, State University of Utrecht.
9. J.A. BERGSTRA (1985). *Put and Get, Primitives for Synchronous Unreliable Message Passing*, Logic Group Preprint Series Nr. 3, Dept. of Philosophy, State University of Utrecht.
10. J.A. BERGSTRA, J.W. KLOP (1982). *Fixed Point Semantics in Process*

Algebras, MC Report IW 206, Amsterdam.
11. J.A. BERGSTRA, J.W. KLOP (1984). The algebra of recursively defined processes and the algebra of regular processes. J. PAREDAENS (ed.). *Proceedings 11th ICALP, Antwerpen 1984, Springer LNCS 172,* 82-95.
12. J.A. BERGSTRA, J.W. KLOP (1985). Algebra of communicating processes with abstraction. *TCS 37 (1),* 77-121.
13. J.A. BERGSTRA, J.W. KLOP (1984). Verification of an alternating bit protocol by means of process algebra. W. BIBEL, K.P. JANTKE (EDS.). Proc. of the International Spring School, Wendisch.-Rietz (GDR), April 1985, *Math. Methods of Spec. and Synthesis of Software Systems '85,* Akademie-Verlag Berlin 1986.
14. J.A. BERGSTRA, J.W. KLOP (1984). Process algebra for synchronous communication. *Information and Control 60 (1/3),* 109-137.
15. J.A. BERGSTRA, J.W. KLOP (1986). Algebra of communicating processes. J.W. DE BAKKER, M. HAZEWINKEL, J.K. LENSTRA (eds.). *Mathematics and Computer Science, CWI Monograph 1,* North-Holland, Amsterdam.
16. J.A. BERGSTRA, J.W. KLOP, E.-R. OLDEROG (1985). *Readies and Failures in the Algebra of Communicating Processes,* CWI Report CS-R8523, Amsterdam.
17. J.A. BERGSTRA, J.W. KLOP, E.-R. OLDEROG (1986). *Failure Semantics with Fair Abstraction,* CWI Report CS-R8609, Amsterdam.
18. J.A. BERGSTRA, J.W. KLOP, J.V. TUCKER (1983). Algebraic tools for system construction. E. CLARKE, D. KOZEN (eds.). *Logics of Programs, Proceedings '83, Springer LNCS 164,* 34-45.
19. J.A. BERGSTRA, J. TIURYN (1983). *Process Algebra Semantics for Queues,* MC Report IW 241, Amsterdam.
20. J.A. BERGSTRA, J.V. TUCKER (1984). *Top Down Design and the Algebra of Communicating Processes,* Sci. of Comp. Progr. 5 (2), p. 171-199.
21. R.J. VAN GLABBEEK (1986). *Bounded Nondeterminism and the Approximation Induction Principle in Process Algebra,* to appear as CWI Report, Amsterdam.
22. W. HESSELINK (1986). *Morfismen van Procesruimten, Universele Domeinen, Formele Implementaties en Fairness,* manuscript, Univ. of Groningen.
23. C.A.R. HOARE (1978). Communicating sequential processes. *Comm. ACM 21,* 666-677.
24. C.A.R. HOARE (1984). *Notes on Communicating Sequential Processes,* International Summer School in Marktoberdorf: Control Flow and Data Flow, Munich.
25. C.A.R. HOARE (1985). *Communicating Sequential Processes,* Prentice Hall.
26. C.P.J. KOYMANS, J.C. MULDER (1986). *A Modular Approach to Protocol Verification using Process Algebra,* Logic Group Preprint Series Nr. 6, Dept. of Philosophy, State University of Utrecht.
27. C.P.J. KOYMANS, J.L.M. VRANCKEN (1985). *Extending Process Algebra with the Empty Process* ε, Logic Group Preprint Series Nr. 1, Dept. of Philosophy, State University of Utrecht.
28. E. KRANAKIS (1986). *Approximating the Projective Model,* CWI Report

CS-R8607, Amsterdam.
29. E. KRANAKIS (1986). *Fixed Point Equations with Parameters in the Projective Model,* CWI Report CS-R8606, Amsterdam.
30. R. MILNER (1980). *A Calculus of Communicating Systems,* Springer LNCS 92.
31. R. MILNER (1984). A complete inference system for a class of regular behaviours. *Journal of Computer and System Sciences 28, 3,* 439-466.
32. D.M.R. PARK (1981). Concurrency and automata on infinite sequences. *Proc. 5th GI Conference, Springer LNCS 104.*
33. F.W. VAANDRAGER (1986). *Verification of Two Communication Protocols by Means of Process Algebra,* CWI Report CS-R8608, Amsterdam.
34. J.L.M. VRANCKEN (1986). *The Algebra of Communicating Processes with Empty Process,* Report FVI 86-01, Computer Science Department, University of Amsterdam.

Codes from Algebraic Number Fields

H.W. Lenstra, Jr.
Universiteit van Amsterdam, Mathematisch Instituut
P.O. Box 19268, 1000 GG Amsterdam, The Netherlands

INTRODUCTION

The geometry of numbers, coding theory, the Riemann hypothesis - the list of *key words* for this lecture can be read as a partial history of the *Stichting Mathematisch Centrum*. The lecture itself attempts to reflect the spirit of the *SMC* by displaying a new connection between these subjects. Using ideas from the *geometry of numbers* one can construct a class of *codes* from algebraic number fields, and the study of the asymptotic properties of these codes depends on the *generalized Riemann hypothesis*.

The construction described in this lecture is a generalization to algebraic number fields of the following idea to make a code. Let P be a finite set of prime numbers, and consider, for a suitable positive integer k, the set C of all elements

$$c_i = (i \bmod p)_{p \in P} \in \prod_{p \in P} \mathbb{Z}/p\mathbb{Z}, \quad i = 1, 2, \ldots, k.$$

If, for $i > j$, the elements c_i, c_j of this set agree on many coordinates then the difference $i - j$ is divisible by many primes, so also by their product. But this difference is less than k, which may lead to a contradiction. This gives us control over the minimum distance of C.

The codes just described have several undesirable properties. First, they are *mixed* codes in the sense that the alphabet size p is not constant. Secondly, they are non-linear, although they are still 'half-linear' in the sense that for any two distinct $x, y \in C$ one of $x - y$, $y - x$ belongs to C. Thirdly, for bounded alphabet size the above construction gives only finitely many codes. This means that the usual 'asymptotic' way of judging the quality of a class of codes, which we discuss in Section 1, does not apply to them. Finally, the codes that we described are in all respects inferior to the codes that are obtained in an analogous way if one replaces the ring $\mathbb{Z}$ by the polynomial ring $\mathbb{F}_q[X]$ in one variable over a suitably chosen finite field $\mathbb{F}_q$, and P by a collection of polynomials of the form $X-\alpha$ with $\alpha \in \mathbb{F}_q$. These codes, the *generalized Reed-Solomon codes* [6, Chapter 10, Section 8], have at least the same minimum distance and dimension, they are linear and non-mixed, but they do have the third shortcoming just mentioned.

If we generalize the construction to algebraic number fields, as we do in Section 2, the situation changes only slightly. For any algebraic number field different from $\mathbb{Q}$ it is true that the ring of integers has different prime ideals with isomorphic residue class fields. Hence it would seem possible to make non-mixed codes by the same recipe. However it turns out that it is better to make non-mixed codes by starting from mixed codes that have a slight variation in the alphabet size. This leaves at least the possibility open to obtain satisfactory asymptotic results (see the remark on $r=q$ at the end of Section 3).

Our codes remain non-linear; even the 'half-linearity' mentioned above disappears.

For fixed alphabet size, the new construction gives infinitely many codes, so that in principle their quality can be analyzed asymptotically. Section 3 contains upper and lower bounds for how good our codes are. These bounds can be substantially improved if one assumes the truth of the generalized Riemann hypothesis, but even then there is a considerable gap between the upper and the lower bound.

The new codes are the analogues, for number fields, of the codes constructed by Goppa and Tsfasman [7, 12] from curves over finite fields. For the analogy between number fields and curves over finite fields, see [1, 14]. If the generalized Riemann hypothesis is true our codes are, asymptotically speaking, not as good as those of Goppa and Tsfasman. Also, the latter codes are linear and non-mixed.

We finally note that there is a non-constructive element in the description of our codes, so that it is still too early to ask for encoding and decoding algorithms. It can be imagined that lattice basis reduction algorithms [5] play a role in this context.

1. CODES

In this section we follow MANIN [7, Section 2], except that we do not require codes to be linear.

Let q be an integer, $q>1$, and V a set of cardinality q, to be referred to as the *alphabet*. For each integer $n \geq 0$ we define a metric w on the set V^n by letting $w(x,y)$ be the number of coordinates where x and y differ. A *code* over V is a non-empty set C that for some integer $n \geq 0$ is a subset of V^n. The number n is called the *word length* of the code. The *dimension* $\dim(C)$ of the code is defined to be $(\log \# C)/(\log q)$, where $\#$ denotes cardinality. The *minimum distance* or simply *distance* $d(C)$ of the code C is the minimum of the numbers $w(x,y)$ if (x,y) runs over all pairs of distinct elements of C; for $\#C=1$ this is $+\infty$.

We are interested in finding codes for which the dimension and the distance are large as functions of the word length. Each code C of positive word length n and positive dimension gives rise to a point $(d(C)/n, \dim(C)/n)$ of the unit square $[0,1]^2$. If C runs over all such codes we obtain a sequence of points in the unit square, and we denote by U_q the set of limit points of this sequence. (If q is a prime power, this set contains the corresponding set from [7].)

As in [7] we have the following result (but not necessarily with the same function α_q).

THEOREM (1.1). *There is a continuous function* $\alpha_q : [0,1] \to [0,1]$ *such that*

$$U_q = \{(x,R): 0 \leq x \leq 1, \ 0 \leq R \leq \alpha_q(x)\}.$$

The function α_q assumes the value 1 in $x=0$, is strictly decreasing on the interval $[0,(q-1)/q]$, and vanishes on the interval $[(q-1)/q, 1]$. Moreover, for $0 \leq x \leq (q-1)/q$ one has

$$\beta_q(x) \leq \alpha_q(x) \leq 1 - \frac{q}{q-1}x$$

where

$$\beta_q(x) = 1 - \frac{x\log(q-1) - x\log x - (1-x)\log(1-x)}{\log q}.$$

PROOF (sketch). It is easy to make codes that show that the points in the unit square that lie on the coordinate axes belong to U_q. Next let $(x,R) \in U_q$. Trivial constructions on codes, such as omitting code words or changing letters, show that the rectangle $[0,x] \times [0,R]$ is contained in U_q. Other constructions, such as projecting a code $C \subset V^n$ to V^{n-1} or intersecting it with a suitably embedded $V^{n-1} \subset V^n$, show that the line segments connecting (x,R) with $(0, R/(1-x))$ and $(x/(1-R), 0)$ are contained in U_q. (These line segments form part of the lines connecting (x,R) with $(1,0)$ and $(0,1)$.)

These results imply that U_q can be described, as in the theorem, by means of a non-increasing function α_q, that α_q is continuous except possibly at 0, and that it is strictly decreasing on the interval where it does not vanish.

The Plotkin bound [13, Theorem 5.2.5] implies that α_q vanishes on $[(q-1)/q, 1]$, and by the above results this leads to the upper bound stated in the theorem. The lower bound is the Gilbert-Varshamov bound [13, Theorem 5.1.9]. It implies continuity of α_q at $x=0$.

This concludes the proof of the theorem.

For better upper bounds on α_q we refer to [13, Chapter 5]. Only recently a better lower bound was found, and only for relatively large q. This was done with the help of modular curves and Shimura curves over finite fields [12].

The following result is useful in comparing the asymptotic properties of the codes that we shall construct with the Gilbert-Varshamov bound.

PROPOSITION (1.2). *Let $r \in \mathbb{R}$, $r \geq 1$. Then the line*
$$R = (1-x)\frac{\log r}{\log q} - \frac{\log((q+r-1)/q)}{\log q}$$
is tangent to the graph of β_q at the point (x_0, R_0), where
$$x_0 = \frac{q-1}{q+r-1},$$
$$R_0 = \frac{r \log r}{(q+r-1)\log q} - \frac{\log((q+r-1)/q)}{\log q}.$$

The proof is straightforward.

2. NUMBER FIELDS

Let K be a *number field*, i.e. a field that is of finite degree m over the field $\mathbb{Q}$ of rational numbers, and let $s, t \in \mathbb{Z}$ be such that there is an isomorphism $K \otimes_{\mathbb{Q}} \mathbb{R} \cong \mathbb{R}^s \times \mathbb{C}^t$ of $\mathbb{R}$-algebras. Denote by A the ring of integers of K, and by Δ the absolute value of its discriminant over $\mathbb{Z}$. The *norm* $\mathfrak{N}(\mathfrak{p})$ of a non-zero prime ideal $\mathfrak{p}$ of A is the cardinality of its residue class field $A/\mathfrak{p}$. For background on algebraic number theory we refer to [2, 11].

THEOREM (2.1). *Let K be a number field, and s, t, Δ as above. Let r, q be integers satisfying $1 < r \leq q$, and write*
$$n = s + t + \#\{\mathfrak{p}: r \leq \mathfrak{N}(\mathfrak{p})^{k(\mathfrak{p})} \leq q \text{ for some } k(\mathfrak{p}) \in \mathbb{Z}_{>0}\};$$
here $\mathfrak{p}$ ranges over the non-zero prime ideals of the ring of integers of K. Then for any positive integer d there exists a code of word length n over an alphabet of q letters with distance at least d and dimension at least
$$(n+1-d)\frac{\log r}{\log q} - \frac{\log\sqrt{\Delta}}{\log q}.$$

PROOF. Let it first be assumed that K is *totally real*, i.e. $s=m$, $t=0$. Under the embedding $K \subset K \otimes_{\mathbb{Q}} \mathbb{R} \cong \mathbb{R}^m$ the ring A becomes a lattice, and if F denotes a fundamental domain for A then F has volume $\text{vol}(F) = \sqrt{\Delta}$.

Let U be the set of those $(x_i)_{i=1}^m \in \mathbb{R}^m$ for which
$$0 < x_i < r^{(n+1-d)/m} \text{ for } 1 \leq i \leq m.$$
This is an open subset of $\mathbb{R}^m$, and $\text{vol}(U) = r^{n+1-d}$.

In analogy with the construction mentioned in the introduction one would now be inclined to make a code from the set $U \cap A$. A basic principle of the geometry of numbers suggests that $\#U \cap A$ is approximately equal to $\text{vol}(U)/\sqrt{\Delta}$, but it turns out that the error term may dominate. To solve this problem we average over all translations of U, which is a 'non-constructive' element in the description of the code.

Let χ denote the characteristic function of U. We have:

$$\int_{z\in F} \#((z+U)\cap A)\,dz = \sum_{y\in A}\int_{z\in F}\chi(y-z)\,dz = \sum_{y\in A}\int_{z\in y-F}\chi(z)\,dz$$

$$= \int_{z\in \mathbb{R}^m}\chi(z)\,dz = \text{vol}(U) = \int_{z\in F}\frac{\text{vol}(U)}{\sqrt{\Delta}}dz,$$

where we use that $\mathbb{R}^m$ is the disjoint union of the sets $y-F$, $y\in A$. It follows that there exists $z\in F$ with $\#((z+U)\cap A)\geqslant \text{vol}(U)/\sqrt{\Delta}$. Let such a z be chosen, and put $C = (z+U)\cap A$. Then we have

$$\#C \geqslant \frac{\text{vol}(U)}{\sqrt{\Delta}} = \frac{r^{n+1-d}}{\sqrt{\Delta}}.$$

Let $V = \{0,1,\cdots,q-1\}$. For each $j\in\{1,2,\cdots,m\}$ we define a map $z+U\to V$ by dividing the projection of $z+U$ on the j-th coordinate axis into q intervals of equal length; i.e., the point $z+(x_i)_{i=1}^m \in z+U$ is mapped to $v\in V$ if

$$\frac{vr^{(n+1-d)/m}}{q} \leqslant x_j < \frac{(v+1)r^{(n+1-d)/m}}{q}.$$

Restricting this map to C we obtain a map $f_j:C\to V$.

For each $\mathfrak{p}$ as in the definition of n choose a positive integer $k(\mathfrak{p})$ with $r\leqslant \mathfrak{N}(\mathfrak{p})^{k(\mathfrak{p})}\leqslant q$ and an injective map $A/\mathfrak{p}^{k(\mathfrak{p})}\to V$. Let $f_\mathfrak{p}:C\to V$ be the composed map $C\subset A\to A/\mathfrak{p}^{k(\mathfrak{p})}\to V$.

Combining all maps f_j, $f_\mathfrak{p}$ we obtain a map $f:C\to V^n$. We claim that

$$\text{if } x,y\in C,\ x\neq y,\ \text{then } w(f(x),f(y))\geqslant d$$

where w denotes the Hamming distance (see Section 1). To prove this, let a be the number of j's for which $f_j(x)=f_j(y)$ and b the number of $\mathfrak{p}$'s for which $f_\mathfrak{p}(x)=f_\mathfrak{p}(y)$, so that $a+b=n-w(f(x),f(y))$. Denote by $N:K\to\mathbb{Q}$ the absolute value of the norm function. We estimate $N(x-y)$ in two ways. On the one hand, all conjugates of $x-y$ are less than $r^{(n+1-d)/m}$ in absolute value, and a of them are even a factor q smaller, so

$$N(x-y) < r^{n+1-d}/q^a \leqslant r^{n+1-d-a}.$$

On the other hand, $x-y$ is a non-zero algebraic integer belonging to b of the ideals $\mathfrak{p}^{k(\mathfrak{p})}$, which each have norm at least r, so that

$$N(x-y) \geqslant r^b.$$

It follows that $b<n+1-d-a$, so $w(f(x),f(y))=n-a-b\geqslant d$. This proves the claim.

It follows in particular that f is injective. Hence the code $f[C]\subset V^n$ has dimension $(\log\#C)/(\log q)$, which is at least $((n+1-d)\log r - \log\sqrt{\Delta})/(\log q)$. By the claim, the distance of $f[C]$ is at least d.

This proves the theorem in the case that K is totally real. To deal with the general case in the same way one needs an analogue, in the complex plane, of a real interval that is divided into q intervals that are q times as small. More precisely, one needs the following result.

For every positive integer q there exists a subset of the euclidean plane that has area $q/2$ and diameter $\leq \sqrt{q}$, and that can be written as the union of q sets of diameter ≤ 1.

If q is a square this is proved by subdividing a square of area $q/2$ into q squares in the obvious way. We leave the elementary proof of the general case to the reader. The result can actually be improved, which gives rise to a slightly better lower bound for the dimension of the code, if $t > 0$.

This completes the proof of the theorem.

To describe the asymptotic properties of the codes from Theorem (2.1) we introduce the following quantity. Let r, q be as in the theorem. Then we define

$$A(q,r) = \liminf_{K} \frac{\log \sqrt{\Delta}}{n \log q},$$

the liminf ranging over all number fields K, up to isomorphism, with n, Δ as in the theorem.

COROLLARY (2.2). *The segment of the line*

$$R = (1-x)\frac{\log r}{\log q} - A(q,r)$$

for which $0 \leq x$, $R \leq 1$ lies entirely in the code domain U_q.

This is an immediate consequence of Theorem (2.1) and the results of Section 1.

3. ASYMPTOTICS

Let $A(q,r)$ be as defined in Section 2.

PROPOSITION (3.1). *There are positive constants c_1, c_2 such that $A(q,r) \geq c_1/q$ and $A(q,q) \geq c_2/\log q$ for all integers r, q with $1 < r \leq q$.*

PROOF. For a number field K, let m, Δ, n be as in Section 2. Known lower bounds for discriminants (see [9]) imply that there is a positive constant c_3 such that $\log \Delta \geq c_3 m$ for all $K \neq \mathbf{Q}$. Moreover, it is obvious that $n \leq m \cdot (1 + \pi(q))$, where $\pi(q)$ denotes the number of prime numbers $\leq q$, and that $n \leq 2m$ if $r = q$. Since $\pi(q) \leq c_4 q / \log q$ for some positive constant c_4 and all q, the proposition follows. This proves (3.1).

It is amusing to note that the first inequality of (3.1) can also be deduced from (1.1) and (2.2), as follows. It is easy to see that n is maximal if r is the least integer $\geq \sqrt{q}$; so let this be the case. Putting $x = (q-1)/q$ in (2.2) and using that α_q vanishes in this point one finds that

$$\frac{1}{q} \cdot \frac{\log r}{\log q} - A(q,r) \leq 0$$

so $A(q,r) \geq 1/(2q)$, as required.

The second inequality of (3.1) is best possible, apart from the value of the constant, as we shall see in (3.4). The first inequality of (3.1) can be sharpened if we assume the *generalized Riemann hypothesis*:

(GRH) *for every number field K, the Dedekind zeta function ζ_K has no complex zeroes with real part larger than 1/2.*

PROPOSITION (3.2). *Let for every integer $q>1$ and every number field K the quantity $B_q(K)$ be defined by*

$$B_q(K) = (\frac{1}{2}(\log 8\pi + \gamma + \frac{\pi}{2})s + (\log 8\pi + \gamma)t + \sum_{\mathfrak{p}, \mathfrak{N}(\mathfrak{p}) \leq q} \frac{\log \mathfrak{N}(\mathfrak{p})}{\sqrt{\mathfrak{N}(\mathfrak{p})} - 1}) / \log \sqrt{\Delta}.$$

Here the summation ranges over non-zero prime ideals $\mathfrak{p}$ of the ring of integers of K, and $s, t, \Delta, \mathfrak{N}(\mathfrak{p})$ are as in Section 2. Further, γ denotes Euler's constant. Suppose moreover that (GRH) is true. Then for every integer $q>1$ we have

$$\limsup_{K} B_q(K) \leq 1,$$

the limsup *ranging over all number fields K, up to isomorphism.*

PROOF. This is an easy consequence of Weil's 'explicit formulae' in the theory of prime numbers, cf. [9, 10, 4]. This proves (3.2).

PROPOSITION (3.3). *Assume (GRH). Then for all integers r,q with $1<r\leq q$ one has*

$$A(q,r) > \frac{1}{\sqrt{q} - 1}.$$

PROOF. This follows from (3.2) by a direct calculation. This proves (3.3).

Next I consider upper bounds.

PROPOSITION (3.4). *There is a positive constant c_5 such that $A(q,r) \leq c_5/\log q$ for all integers r, q with $1<r\leq q$.*

PROOF. By the theory of infinite class field towers [2, Chapter IX] there exists a number field E such that the maximal totally unramified extension L of E is of infinite degree over E. We let K range over the finite extensions of E that are contained in L, and for each K we let m, Δ, s, t, n be as in Section 2. Each K is unramified over E, so the number $\Delta^{1/m}$ is independent of K. Also, one has $n \geq s+t \geq m/2$. It follows that

$$\liminf_{K, E \subset K \subset L} \frac{\log \sqrt{\Delta}}{n \log q} \leq \frac{c_5}{\log q}$$

for some positive constant c_5. This proves (3.4).

By (3.1), the inequality of (3.4) is best possible for $r=q$, apart from the value of the constant. If r is much smaller than q we can again use the generalized Riemann hypothesis to obtain a better result. For the sake of definiteness I choose r to be the least integer $\geq q/2$.

PROPOSITION (3.5). *Suppose that* (GRH) *is true. Then there is a positive constant* c_6 *such that for every integer* $q>1$ *we have* $A(q,[(q+1)/2]) \leq c_6(\log q)/q^{1/4}$.

The proof depends on two lemmas.

LEMMA (3.6). *Suppose that q is an integer, $q>1$, and that k,l are positive integers. Write*
$$d = 4 \cdot \prod_{i=1}^{l} p_i,$$
where p_i denotes the i-th prime number. Suppose that the following two conditions are satisfied.

(i) $k+1 \leq \frac{1}{4}(l-1)^2 - (l-1)$;

(ii) *there are at least k prime numbers p with $\sqrt{q/2} \leq p \leq \sqrt{q}$ for which the Legendre symbol $(\frac{-d}{p})$ equals -1.*

Then we have
$$A(q, [(q+1)/2]) \leq \frac{\log d}{2(k+1)\log q}.$$

PROOF. Let E be the imaginary quadratic field with discriminant $-d$. Each p as in (ii) generates a principal prime ideal $\mathfrak{p}$ of the ring of integers of E with $q/2 \leq \mathfrak{N}(\mathfrak{p}) \leq q$. Let S be a set of k such prime ideals. Denote by L the maximal totally unramified extension of E in which all $\mathfrak{p} \in S$ split completely. Using a slight generalization of the theory of infinite class field towers (see [4, Section 14]) one deduces from inequality (i) that L is of infinite degree over E. (Since all $\mathfrak{p} \in S$ are principal, the number t from [4, Section 14] equals $l-1$, and $\rho = k+1$.) To prove the lemma, let now K range over the finite extensions of E that are contained in L, as in the proof of (3.4). As before, the number $\Delta^{1/m}$ is independent of K, and putting $K=E$ one sees that it equals $\sqrt{d}$. Also, since each $\mathfrak{p} \in S$ splits completely in K one has $n \geq t + [K:E] \cdot \#S = \frac{1}{2}m(1+k)$ for each K. This proves (3.6).

LEMMA (3.7). *Assume* (GRH). *Then for every positive real number c_7 there is a positive real number c_8 with the following property.*

Let d be a positive integer for which $-d$ is the discriminant of a quadratic field. Then for every real number x with $x \geq c_8 (\log d)^2$ the number of odd prime numbers p for which

$$x / \sqrt{2} \leq p \leq x, \quad \left(\frac{-d}{p}\right) = -1$$

is at least $c_7 (\log d)^2 / \log\log d$.

PROOF. This is proved by a slight adaptation of the proof of [8, Theorem 13.1; pp. 120, 123, 124] (the weight function $(1 - n/N)$ should be changed so as to count primes in the right interval). I thank H.L. MONTGOMERY for pointing this out to me.

PROOF OF (3.5). For any integer $l \geq 7$, let $k = k(l)$ be the largest integer satisfying (3.6)(i), and let $d = d(l)$ be as in (3.6). Then we have

$$\log d \sim l \cdot \log l, \quad k \sim (1/4)(\log d)^2 / (\log\log d)^2$$

for $l \to \infty$, so there is certainly a positive constant c_7 such that $k \leq c_7 (\log d)^2 / \log\log d$ for all $l \geq 7$. Let c_8 be the number that Lemma (3.7) guarantees to exist.

Now let q be an integer, $q > 1$, and choose the integer l as large as possible subject to the condition $\sqrt{q} \geq c_8 (\log d(l))^2$. We suppose that q is sufficiently large for l to be well-defined and ≥ 7. By the choice of c_7 and Lemma (3.7), the conditions of (3.6) are satisfied for $k = k(l)$ and l, so (3.6) gives us an upper bound on $A(q, [(q+1)/2])$. We have

$$\log d \sim c_9 q^{1/4}, \quad k \sim (1/4)(\log d)^2 / (\log\log d)^2 \sim c_{10} \sqrt{q} / (\log q)^2$$

for certain positive constants c_9, c_{10}, as $q \to \infty$. It follows that the upper bound from (3.6) leads to the upper bound stated in Proposition (3.5), at least for q sufficiently large. For the remaining values of q one can apply (3.4). This proves (3.5).

We discuss the implications of our estimates for coding theory.

The first inequality of (3.1) yields a rather crude upper bound for how good we can expect our codes to be. I do not know how this bound compares to the best upper bounds that are known for the function α_q of Section 1. It is conceivable that these, together with (2.2), lead to a better lower bound for $A(q,r)$.

The second inequality of (3.1) shows that it is not advisable to apply our construction only with $r = q$. By (1.2) that would at best lead to codes that are comparable to the codes realizing the Gilbert-Varshamov bound.

Proposition (3.3) is the analogue of the result that was proved by DRINFELD and VLADUT [3] for function fields of curves over finite fields. For very small q, such as $q = 2$, it shows that one should not expect our codes to lead to a point (x, R) of the code domain U_q with x and R positive. For large q, Propositions (3.3) and (1.2) show that one can still hope to find codes that beat the Gilbert-Varshamov bound. In the case of function fields this hope was indeed realized for certain values of q, see [12].

It is apparently harder to construct good codes from number fields. Proposition (3.4) leads to codes whose performance is comparable to the Gilbert-Varshamov bound. Proposition (3.5) shows that much better codes can be made, for large q, if one again accepts (GRH). However, these codes are not as good as those made with function fields, and there remains a substantial gap between the bounds of (3.3) and (3.5). The analogy with function fields suggests that (3.3) is nearer to the truth than (3.5).

REFERENCES

1. E. ARTIN, G. WHAPLES (1945). Axiomatic characterization of fields by the product formula for valuations. *Bull. Amer. Math. Soc. 51*, 469-492; pp. 202-225 in: *The Collected Papers of Emil Artin*, Addison-Wesley Publishing Company, Reading 1965.
2. J.W.S. CASSELS, A. FRÖHLICH (eds.) (1967). *Algebraic Number Theory*, Academic Press, London.
3. V.G. DRINFELD, S.G. VLADUT (1983). On the number of points on an algebraic curve (in Russian). *Funktsional. Anal. i. Prilozhen. 17 (1)*, 68-69; English translation: *Functional Anal. Appl. 17*, 53-54.
4. Y. IHARA (1983). How many primes decompose completely in an infinite unramified Galois extension of a global field? *J. Math. Soc. Japan 35*, 693-709.
5. A.K. LENSTRA, H.W. LENSTRA, JR., L. LOVÁSZ (1982). Factoring polynomials with rational coefficients. *Math. Ann. 261*, 515-534.
6. F.J. MACWILLIAMS, N.J.A. SLOANE (1978). *The Theory of Error-Correcting Codes*, North-Holland Publishing Company, Amsterdam.
7. YU.I. MANIN, (1981). What is the maximum number of points on a curve over $\mathbb{F}_2$? *J. Fac. Sci. Univ. Tokyo Sect. IA Math. 28*, 715-720.
8. H.L. MONTGOMERY (1971). *Topics in Multiplicative Number Theory*, Lecture Notes in Math. 227, Springer-Verlag, Heidelberg.
9. G. POITOU (1977). Minorations de discriminants (d'après A.M. Odlyzko). *Séminaire Bourbaki 28 (1975/76), no. 479*, pp. 136-153 in: *Lecture Notes in Math. 567*, Springer-Verlag, Heidelberg.
10. G. POITOU. Sur les petits discriminants. *Séminaire Delange-Pisot-Poitou (Théorie des nombres) 18 (1976/77), no. 6*.
11. P. SAMUEL (1967). *Théorie Algébrique des Nombres*, Hermann, Paris; English translation: *Algebraic Theory of Numbers*, Houghton Mifflin Company, Boston 1970.
12. M.A. TSFASMAN, S.G. VLADUT, TH. ZINK (1982). Modular curves, Shimura curves, and Goppa codes, better than Varshamov-Gilbert bound. *Math. Nachr. 109*, 21-28.
13. J.H. VAN LINT (1982). *Introduction to Coding Theory*, Graduate Texts in Math. 86, Springer-Verlag, New York.
14. A. WEIL (1939). Sur l'analogie entre les corps de nombres algébriques et les corps de fonctions algébriques. *Revue Scient. 77*, 104-106; pp. 236-240 in: *Oeuvres Scientifiques / Collected Papers*, vol. I, Springer-Verlag, New York 1979.

Infinite-Dimensional Normed Linear Spaces

and

Domain Invariance

Jan van Mill

Vrije Universiteit, Subfaculteit Wiskunde en Informatica
P.O. Box 7161, 1007 MC Amsterdam, The Netherlands
and
Universiteit van Amsterdam, Mathematisch Instituut
P.O. Box 19268, 1000 GG Amsterdam, The Netherlands

The Brouwer invariance of domain property for Euclidean spaces implies that, for open $U \subseteq \mathbb{R}^n$, every injective map $g: U \to \mathbb{R}^n$ is an open imbedding [2]. It is well-known that this property does not hold for infinite-dimensional linear spaces. Indeed, for any infinite-dimensional normed linear space Y we have the following examples:

EXAMPLE 1 ([1, III theorem 6.3]). There exists a homeomorphism $h: Y \to h(Y)$ onto a non-open subset of Y.

PROOF. Let $h_0: S \to H$ be any homeomorphism from the unit sphere S onto a closed hyperplane H. Then for any $y \in Y \setminus H$, h_0 may be extended to a homeomorphism h of Y onto the non-open set $(H + (-\infty, 1) \cdot y) \cup \{y\}$ with $h(0) = y$. □

EXAMPLE 2 (D.W. CURTIS). There exists a bijective map $g: Y \to Y$ such that $g | Y \setminus K$ is not a homeomorphism for any compact K.

PROOF. Since the unit sphere S is non-compact, there exists a map $\lambda: S \to (0, 1]$ such that $\inf \lambda(S) = 0$. Define $f: Y \to Y$ by the formulas

$$\begin{cases} f(y) = \lambda(\frac{y}{\|y\|}) \cdot y & (y \neq 0), \\ f(0) = 0. \end{cases}$$

Clearly, f is a bijective map of Y, but is not a homeomorphism since $0 \notin \mathrm{int} f(B)$ for any bounded set B. Note that $f | Y \setminus \{0\}$ is a homeomorphism.

Using copies of f on a discrete sequence of open balls in Y, we may construct a bijective map $g: Y \to Y$ such that for any compact $K \subseteq Y$, $g|Y \setminus K$ is not a homeomorphism. In fact, there exists an open $U \subseteq Y \setminus K$ such that for any compact $J \subseteq U$, $g(U \setminus J)$ is non-open. However, there is a dense open $V \subseteq Y$ such that $g|V$ is an open imbedding. □

A space X is a *Baire space* if the intersection of any countable family of dense open subsets of X is dense. A function $f: Y \to Y$ on a linear space has *countable type* if there exists a countable set Z in Y such that for each $y \in Y$,

$$f(y) \in \text{span}(\{y\} \cup Z).$$

If each map $f: Y \to Y$ on a topological linear space has countable type, we say that Y has *countable type for maps*.

Clearly $\mathbb{R}^n$ and, more generally, each $\aleph_0$-dimensional topological linear space (i.e. a topological linear space with a countable Hamel basis) has countable type for maps. Consequently, there exist infinite-dimensional topological linear spaces having countable type for maps. However, not all topological linear spaces have this property, see Example 3.

LEMMA 1. *Let Y be a topological linear space and let $A: Y \to Y$ be a linear operator with countable type. Then $A = \lambda I + B$, for some scalar λ and linear operator B with $\aleph_0$-dimensional range.*

PROOF. Let $Z \subseteq Y$ be a countable set such that $A(y) \in \text{span}(\{y\} \cup Z)$ for each y. Let $E \subseteq Y$ be a complementary linear subspace for span Z, and consider any linearly independent set $\{e_1, e_2\} \subseteq E$. There exist scalars λ_1, λ_2 and λ, and elements $s_1, s_2, s \in \text{span } Z$, such that

$$A(e_1) = \lambda_1 \cdot e_1 + s_1;$$
$$A(e_2) = \lambda_2 \cdot e_2 + s_2; \text{ and}$$
$$A(e_1 + e_2) = \lambda \cdot (e_1 + e_2) + s.$$

Using $A(e_1 + e_2) = A(e_1) + A(e_2)$, we obtain

$$(\lambda - \lambda_1) \cdot e_1 + (\lambda - \lambda_2) \cdot e_2 = s_1 + s_2 - s.$$

Since $E \cap \text{span } Z = \{0\}$, $(\lambda - \lambda_1) \cdot e_1 + (\lambda - \lambda_2) \cdot e_2 = 0$ and since $\{e_1, e_2\}$ is linearly independent, $\lambda_1 = \lambda = \lambda_2$. This implies that for any linearly independent set F in E, $(A - \lambda I)(e) \in \text{span } Z$ for each $e \in F$. It follows that $(A - \lambda I)(E) \subseteq \text{span } Z$. Since $Y = E + \text{span } Z$, and $(A - \lambda I)(\text{span } Z) \subseteq \text{span } Z$, we obtain $A - \lambda I : Y \to \text{span } Z$. □

The following result is well-known.

LEMMA 2. *Let $B: Y \to Y$ be a bounded linear operator on a Baire topological linear space. If B has $\aleph_0$-dimensional range, it has finite-dimensional range.*

PROOF. Write $B(Y)$ as $\bigcup_1^\infty F_n$, where each F_n is a finite-dimensional linear subspace of $B(Y)$. Observing that each F_n is closed in $B(Y)$ and that B is continuous, it follows that each $B^{-1}(F_n)$ is closed in Y. Since clearly $\bigcup_1^\infty B^{-1}(F_n) = Y$ and since Y is a Baire space, one of the $B^{-1}(F_n)$'s has non-empty interior, say $B^{-1}(F_{n_0})$. Every proper closed linear subspace of any topological linear space has empty interior. We conclude that $B^{-1}(F_{n_0}) = Y$. □

A topological linear space Y is said to have *few operators* if every bounded linear operator $A: Y \to Y$ has the form $A = \lambda I + B$, for some scalar λ and some operator B with finite-dimensional range.

In [4] the author constructed an infinite-dimensional pre-Hilbert space with few operators (a Banach space B of uncountable weight such that each bounded linear operator $A: B \to B$ has the form $A = \lambda I + E$, for some scalar λ and some operator E with separable range, was earlier constructed under $V = L$ by SHELAH [7]). Lemmas 1 and 2 yield

THEOREM 1. *Let Y be a Baire topological linear space with countable type for maps. Then Y has few operators;*

and

EXAMPLE 3. *Hilbert space l^2 does not have countable type for maps.*

PROOF. Let $E: l^2 \to l^2$ be a bounded linear operator which is of the form $\lambda I + B$, for some scalar λ and some operator B with finite-dimensional range. If $\lambda = 0$ then $E = B$ which implies that E has finite-dimensional range. Suppose that $\lambda \neq 0$. If $E(x) = 0$ then $x = \frac{1}{\lambda} \cdot B(x)$, which belongs to the range of B. Consequently, in this case the kernel of E is finite-dimensional.

By Theorem 1 and the above remarks it suffices to construct a bounded linear operator $A: l^2 \to l^2$ such that neither the range nor the kernel of A is finite-dimensional. This is a triviality of course. Indeed, define $A: l^2 \to l^2$ by

$$A(x_1, x_2, x_3, \ldots) = (x_1, 0, x_3, 0, \ldots).$$

Then A is clearly as required. □

Since each $\aleph_0$-dimensional topological linear space has countable type for maps and Hilbert space l^2 has not, the question naturally arises whether there are topological linear spaces having countable type for maps but which are not $\aleph_0$-dimensional. In [5] the author proved

THEOREM 2 ([5]). *Each separable Banach space B contains a linear subspace Y such that*
(a) *Y is dense in X;*
(b) *Y is a Baire space; and*
(c) *Y has countable type for maps.*

PROOF (sketch). If B is finite-dimensional then $Y=B$ is as required. Therefore assume that B is infinite-dimensional.

Let $g:A\to B$ be a function defined on a subset of B. A subset P of A is said to be *g-independent* if the following conditions are satisfied:

(1) $g|P$ is injective;

(2) $P\cap g(P)=\varnothing$; and

(3) $P\cup g(P)$ is linearly independent.

Via a standard procedure it is possible to prove that if A is a G_δ-subset of B which contains an uncountable g-independent subset then A contains a g-independent Cantor set.

Now let $\mathcal{H}$ denote the collection of all homeomorphisms $h:K_1\to K_2$ between Cantor sets in B such that K_1 is h-independent. It is possible to construct a linear subspace Y of B with the following property

(*) for each $h\in\mathcal{H}$, there exists $x\in\operatorname{dom} h$ such that $x\in Y$ but $h(x)\notin Y$.

Then Y is as required.

By (*) it easily follows that Y intersects every linearly independent Cantor set in B. Since B is infinite-dimensional, every dense G_δ-subset of B contains a linearly independent Cantor set and consequently intersects Y. This implies that Y is a Baire space.

If Y were not dense then the closure of Y would be a proper closed linear subspace of B which therefore would have to be nowhere dense which is impossible since Y intersects every dense G_δ-subset of B. This proves (a) and (b).

For (c), let $f:Y\to Y$ be a map. Since B is complete, f extends to a map $g:A\to B$, for some G_δ-subset A of B. Suppose that A contains an uncountable g-independent set. Then A contains a g-independent Cantor set K. Then $g|K$ is a member of the collection $\mathcal{H}$, and by (*) there exists $x\in K\cap Y$ such that $g(x)\notin Y$. But this contradicts the fact that $g(x)=f(x)\in Y$. Thus every g-independent subset of A is countable, and in particular, every f-independent set is countable.

By (*) Y cannot contain any linearly independent Cantor set. This easily implies that for every countable set P there exists a countable set $\hat{P}$ such that $f(\operatorname{span} P)\subseteq\operatorname{span}\hat{P}$. Now let $Q\subseteq Y$ be a maximal f-independent set. If $Q=\varnothing$ then $f(y)\in\operatorname{span}\{y\}$ for each $y\neq 0$, and f obviously has countable type. Otherwise, construct a tower (P_i) of countable sets by taking $P_1=Q$ and $P_{n+1}=P_n\cup\hat{P}_n$ for each $n\geqslant 1$. Take $Z=\bigcup_1^\infty P_n$. It can be shown that $f(y)\in\operatorname{span}(\{y\}\cup Z)$ for every $y\in Y$. □

COROLLARY 1 ([4]). *There exists an infinite-dimensional pre-Hilbert space with few operators.*

PROOF. By Theorem 2 there is a dense linear subspace $Y\subseteq l^2$ such that Y is a

Baire space and has countable type for maps. Then Y, being dense in ℓ^2, is clearly infinite-dimensional. Also, Y has few operators by Theorem 1. □

Let Y be a topological linear space. Suppose that, for every injective map $g:U\to Y$ with domain an open subset of Y, there exists a nonempty open $V\subseteq U$ such that $g|V$ is an open imbedding. Then we say that Y has *restricted domain invariance*.

We may suppose that V is dense in U, since the condition can be applied to every restriction $g|W$ to an open nonempty $W\subseteq U$. In addition, for Y a normed linear space, it suffices to verify the condition for every injective map $f:Y\to Y$, since there is an open imbedding of Y into every nonempty open subset.

The reader naturally wonders what the relation is between the title of our paper and the results derived or mentioned so far. This is cleared by the following

THEOREM 3 ([5]). *Let Y be a normed linear space with the Baire property and with countable type for maps. Then Y has restricted domain invariance.*

PROOF (sketch). We may assume Y is infinite-dimensional. By the above remark, we need only to consider an injective map $f:Y\to Y$. Let Z be a countable subset of Y such that $f(y)\in\mathrm{span}(\{y\}\cup Z)$ for each y. There exists a tower (A_n) of compacta such that span A_n is finite-dimensional for each n, and $\bigcup_1^\infty A_n = \mathrm{span}\, Z$. For each n, set

$$Y_n = \{y\in Y | \text{for some } \lambda\in[-n,n], f(y)-\lambda\cdot y\in A_n\}.$$

It is easily seen that each Y_n is closed and that $\bigcup_1^\infty Y_n = Y$. Since Y is a Baire space, some Y_n has nonempty interior and since Y is infinite-dimensional, there exists a nonempty open set $W\subseteq Y_n \setminus \mathrm{span}\, A_n$. For each $w\in W$ there is a unique $\lambda(w)\in[-n,n]$ such that $f(w)-\lambda(w)\cdot w\in A_n$; furthermore, the assignment $w\to\lambda(w)$ is continuous. It is possible to show that there exists a nonempty open $V\subseteq W$ with either $\lambda(V)\subseteq[-n,0)$ or $\lambda(V)\subseteq(0,n]$.

For convenience, assume that $\lambda(V)\subseteq(0,n]$ and take $p\in V$ arbitrarily. We may assume that $f(p)=p$. Let $E=\mathrm{span}(\{p\}\cup A_n)$. By using, among other things, the Brouwer invariance of domain property for E, it can be shown that $f(V)$ is a neighborhood of $f(p)=p$. □

Observe that by Examples 1 and 2, Theorem 3 is 'best possible'.

COROLLARY 2 ([5]). *There exists an infinite-dimensional pre-Hilbert space X such that X is not homeomorphic to $X\times\mathbb{R}$.*

PROOF. By Theorems 2 and 3 there is a dense linear subspace $X\subseteq\ell^2$ having restricted domain invariance. X is clearly infinite-dimensional. We claim that X is as required. To the contrary, assume that $\phi:X\to X\times\mathbb{R}$ is a homeomorphism.

Define $\psi: X \to X$ by

$$\psi(x) = \phi^{-1}(x, 0).$$

Then ψ is clearly an imbedding of X onto a subset of X with empty interior. But this contradicts restricted domain invariance. □

The above result generalizes POL [6] and answers Question LS12 in GEOGHEGAN [3].

COROLLARY 3. *Every separable Banach space contains a dense linear subspace Y such that:*
(a) *Y is a Baire space;*
(b) *Y has restricted domain invariance; and*
(c) *Y has few operators.* □

I am indebted to D.W. CURTIS for many helpful comments.

REFERENCES
1. C. BESSAGA, A. PELCZYŃSKI (1975). *Selected Topics in Infinite-Dimensional Topology*, PWN, Warsaw.
2. L.E.J. BROUWER (1912/1913). Invarianz des n-dimensionalen Gebiets. *Math. Ann. 71*, 305-313; *72*, 55-56.
3. R. GEOGHEGAN (1979). Open problems in infinite-dimensional topology. *Top. Proc. 4*, 287-338.
4. J. VAN MILL. An infinite-dimensional pre-Hilbert space all bounded linear operators of which are simple, to appear in *Coll. Math.*
5. J. VAN MILL. Domain invariance in infinite-dimensional linear spaces, to appear in *Proc. Am. Math. Soc.*
6. R. POL (1984). An infinite-dimensional pre-Hilbert space not homeomorphic to its own square. *Proc. Am. Math. Soc. 90*, 450-454.
7. S. SHELAH (1978). A Banach space with few operators. *Isr. J. Math. 30*, 181-191.

Geometric Methods in Discrete Optimization

A. Schrijver

Department of Econometrics Tilburg University
P.O. Box 90153, 5000 LE Tilburg, The Netherlands
and
Centre for Mathematics and Computer Science
P.O. Box 4079, 1009 AB Amsterdam, The Netherlands

1. INTRODUCTION

Classically, there exists a strong connection between optimization and geometry. Often, a set of options ('feasible solutions') can be represented by vectors in euclidean space, and a search process for an optimal option ('option solution') can be seen as a trip in space. The geometric nature of optimization methods like the simplex method, the gradient method, the ellipsoid method, the cutting plane method, is suggested already by their names. Thus optimization illustrates once more that Descartes' idea of analytic geometry can be used in turn to study analytic problems geometrically.

The above being well-known for linear and nonlinear optimization, where the feasible solutions generally give a continuous, sometimes even convex, region in space, the purpose of this paper is to show the geometric character also of several methods and results in combinatorial optimization, where the feasible solutions, in the first instance, yield a discrete, discontinuous set.

Among the geometric methods and results used in combinatorial optimization we discuss are:
- the representation of combinatorial optimization problems by polyhedra;
- the ellipsoid method;
- the basis reduction method for lattices;
- the cutting plane method;
- the results of Tutte and Seymour on the representation and decomposition of geometric configurations in projective spaces over GF(2).

We illustrate these ideas by some applications - we focus on two problems (the matching problem and the coclique problem), but the methods have a much wider applicability (like to trees (directed or undirected), one- and multicommodity flows, coverings, directed cuts, cliques, disjoint paths in graphs, the traveling salesman problem, the acyclic subgraph problem).

2. REPRESENTING COMBINATORIAL OPTIMIZATION PROBLEMS BY POLYHEDRA

The idea of using polyhedra in combinatorial optimization is simple. Suppose we have a collection $\mathcal{F}$ of subsets of a finite set S. (For instance, $\mathcal{F}$ is the collection of matchings in a given graph $G=(V,E)$.) Moreover, a function $c:s\to\mathbb{Z}$ is given, and we wish to find

$$\max_{U\in\mathcal{F}} \sum_{s\in U} c(s), \qquad (1)$$

being a generic form of a combinatorial optimization problem. (In the example above, it amounts to finding a matching of maximum 'weight'.) Usually, the collection $\mathcal{F}$ is too large to evaluate every U in $\mathcal{F}$ to determine the maximum. 'Too large' here means with respect to the data structure given (like the number of matchings in a graph is exponentially large in the size of the graph). One should find a method more efficient than this 'brute force' method.

We can represent each subset U of S by its characteristic vector χ^U in $\{0,1\}^S$, i.e., $(\chi^U)_s = 1$ if $s\in U$, and 0 otherwise. Moreover, the function c can be considered as a vector in $\mathbb{R}^S$. Then problem (1) becomes:

$$\max\{c^T\chi^U | U\in\mathcal{F}\}. \qquad (2)$$

Clearly, the maximum value in (2) is equal to

$$\max\{c^T x | x \in \text{conv.hull}\{\chi^U | U\in\mathcal{F}\}\}, \qquad (3)$$

where conv.hull denotes the convex hull in $\mathbb{R}^S$. Since conv.hull $\{\chi^U | U\in\mathcal{F}\}$ is a convex polytope, there exists a matrix A and a column vector b such that this polytope is equal to $\{x|Ax\leq b\}$ (where the columns of A are indexed by the elements of S). This implies that (3) is equal to

$$\max\{c^T x | Ax\leq b\}. \qquad (4)$$

This way we have transformed the combinatorial optimization problem (1) into a linear programming problem, and we can appeal to linear programming methods to solve the combinatorial problem. We could use the simplex method to solve (4) and hence (1) (note that the simplex method gives a vertex of $\{x|Ax\leq b\}$ as optimal solution, which corresponds to the optimal set in $\mathcal{F}$). Alternatively, one could apply the ellipsoid method for linear programming, which is not a practical method, but which can yield that (1) is solvable in polynomial time.

The mathematical problem now is to determine A and b, given $\mathcal{F}$. Although the system $Ax\leq b$ clearly always exists, there is the problem that in many cases the polytope conv.hull $\{\chi^U | U\in\mathcal{F}\}$ has an enormous number of facets, often too difficult to describe. The application of linear programming methods will be helpful only in case the system $Ax\leq b$ is decent enough - decent in the sense to be described in Section 3 below.

As we shall see also in Section 3, if we are interested in the polynomial-time solvability of combinatorial optimization problems of type (1) (and in fact, we are), the above approach of replacing $\mathcal{F}$ by conv.hull $\{\chi^U | U \in \mathcal{F}\}$ is, at least implicitly, unavoidable.

As a theoretical by-product, if we have written (1) as the LP-problem (4), we can apply the Duality theorem of linear programming to (4), saying:

$$\max\{c^T x | Ax \leq b\} = \min\{y^T b | y \geq 0, y^T A = c^T\}. \tag{5}$$

Therefore,

$$\max_{U \in \mathcal{F}} \sum_{s \in U} c(s) = \min\{y^T b | y \geq 0, y^T A = c^T\}, \tag{6}$$

which is a *min-max relation* for the combinatorial problem. If we can prove that the minimum in (6) has an integer solution, we obtain a purely combinatorial min-max relation.

3. APPLICATION OF THE ELLIPSOID METHOD

The ellipsoid method was shown by KHACHIYAN [15] to solve linear programming problems in polynomial time. In this section we discuss an application of the ellipsoid method to combinatorial optimization.

An (*undirected*) *graph* is a pair $G = (V, E)$, where V is a finite set and E is a collection of unordered pairs from V. The elements of V and E are called *vertices* and *edges*, respectively.

Suppose we are given, for each graph $G = (V, E)$, a collection $\mathcal{F}_G$ of subsets of E. For example:

(i) $\mathcal{F}_G$ is the collection of matchings in G (a *matching* is a collec-   (7)
tion of pairwise disjoint edges);
(ii) $\mathcal{F}_G$ is the collection of trees in G (a *tree* is a connected set of edges not containing a circuit);
(iii) $\mathcal{F}_G$ is the collection of Hamilton circuits in G (a *Hamilton circuit* is a circuit containing each vertex of G exactly once).

With the family $(\mathcal{F}_G | G \text{ graph})$ we can associate the following problem:

Optimization problem: Given a graph $G = (V, E)$ and $c \in \mathbb{Q}^E$,   (8)
find $E' \in \mathcal{F}_G$ maximizing $\sum_{e \in E'} c(e)$.

So if $(\mathcal{F}_G | G \text{ graph})$ is as in (i), (ii) and (iii), problem (8) amounts to the problems of finding a maximum weighted matching, a maximum weighted tree, and a maximum weighted Hamilton circuit, respectively. The last problem is the well-known traveling salesman problem (note that by replacing c by $-c$, (8) becomes a minimization problem).

Clearly, for each collection $(\mathcal{F}_G | G \text{ graph})$, problem (8) forms a class of problems of type (1).

Especially, we are interested for which families $(\mathcal{F}_G | G \text{ graph})$ problem (8) is solvable in *polynomial time* (or *polynomially solvable*), i.e., solvable by an algorithm whose running time is bounded by a polynomial in the input size, which is

$$|V| + |E| + \text{size}(c). \tag{9}$$

Here $\text{size}(c) := \Sigma_{e \in E} \text{size}(c(e))$, where the size of a rational number p/q is $\log_2(|p|+1) + \log_2(|q|)$. So size (c) is about the space needed to specify c in binary notation.

If $(\mathcal{F}_G | G \text{ graph})$ is as in (7) (i) or (ii), problem (8) is polynomially solvable. If it is as in (iii) no polynomial-time algorithm has been found, and it is a general belief that no such algorithm exists (see also the Remark below).

It has been shown by GRÖTSCHEL, LOVÁSZ and SCHRIJVER [11] that, for any fixed family $(\mathcal{F}_G | G \text{ graph})$, (8) is polynomially solvable, if and only if the following problem is solvable in polynomial time:

Separation problem: Given a graph $G = (V, E)$ and $x \in \mathbb{Q}^E$, determine if x belongs to conv.hull $\{\chi^F | F \in \mathcal{F}_G\}$, and if not, find a separating hyperplane. (10)

Again 'polynomial-time' means: with running time bounded by a polynomial in $|V| + |E| + \Sigma_{e \in E} \text{size}(x(e))$.

THEOREM 1. *For any collection $(\mathcal{F}_G | G \text{ graph})$, the optimization problem (8) is polynomially solvable, if and only if the separation problem (10) is polynomially solvable.*

The theorem implies that with respect to the question of polynomial-time solvability, the approach described in Section 2 (studying the convex hull) is more or less essential: a combinatorial optimization problem is polynomially solvable if and only if the corresponding convex hull can be decently described in the sense of the polynomial solvability of the separation problem. This can also be used in the negative: if a combinatorial optimization problem is not polynomially solvable (maybe the traveling salesman problem), then the corresponding polytopes have no decent description.

Theorem 1 is proved with the help of the ellipsoid method, for which we refer to the books by GRÖTSCHEL, LOVÁSZ and SCHRIJVER [12] and SCHRIJVER [22]. The ellipsoid method does not give practical algorithms, but in some cases with Theorem 1 the polynomial solvability of a combinatorial optimization problem was proved, which next formed a motivation for finding a practical polynomial-time algorithm for the problem.

There are several variations of Theorem 1. For instance, a similar result holds if we consider collections $\mathcal{F}_G$ of subsets of the vertex set V, instead of subsets of the edge set E. E.g., we could take:

$\mathcal{F}_G$ is the collection of all cocliques of G (a *coclique* is a set of vertices which are pairwise not adjacent). (11)

Moreover, a similar theorem holds if we consider classes $(\mathcal{F}_G | G \in \mathcal{G})$, where $\mathcal{G}$ is a subcollection of the set of all graphs. Similarly, we can consider *directed* graphs.

REMARK. The question $NP = P$? amounts to the following. Call a class ($\mathcal{F}_G|G$ graph) *polynomially recognizable* if there exists a polynomial-time algorithm for the following problem:

$$\text{given } G = (V,E) \text{ and } F \subseteq E, \text{ decide if } F \text{ belongs to } \mathcal{F}_G. \tag{12}$$

It is not difficult to see that each of the examples in (7) gives a polynomially recognizable class.

Now one has:

$NP = P$, if and only if for each polynomially recognizable (13) class ($\mathcal{F}_G|G$ graph) the optimization problem is polynomially solvable.

There seems no reason to believe that for every polynomially recognizable class the optimization problem is polynomially solvable. However, no counterexample has been found. It has been shown by COOK [4] and KARP [14] that the traveling salesman problem (and several other classical combinatorial optimization problems) is *NP-complete*. It implies: if the traveling salesman problem is polynomially solvable, then for every polynomially recognizable class the optimization problem is polynomially solvable. This is one of the reasons why a lot of research has been spent on the traveling salesman problem.

4. LATTICE BASIS REDUCTION, SIMULTANEOUS DIOPHANTINE APPROXIMATION AND STRONGLY POLYNOMIAL ALGORITHMS

The basis reduction method for lattices was given by LENSTRA, LENSTRA and LOVÁSZ [16]. It solves the following problem:

given a nonsingular rational $n \times n$-matrix A, find a basis (14)
$b_1, \ldots, b_n$ for the lattice generated by the columns of A satisfying

$$\|b_1\| \cdot \ldots \cdot \|b_n\| \leq 2^{\frac{1}{4}n(n-1)} |\det A|,$$

in time bounded by a polynomial in size(A) $:= \Sigma_{i,j}$ size(a_{ij}). Here the *lattice generated by* $a_1, \ldots, a_n$ is the set of vectors $\lambda_1 a_1 + \ldots + \lambda_n a_n$ with $\lambda_1, \ldots, \lambda_n \in \mathbb{Z}$. Any linearly independent set of vectors generating the lattice is called a *basis* for the lattice.

The basis reduction method has several applications in linear and integer programming, in number theory and in cryptography. One consequence is a polynomial-time algorithm for *simultaneous diophantine approximation*:

THEOREM 2. *There exists a polynomial-time algorithm which for given vector $a \in \mathbb{Q}^n$ and rational ϵ with $0 < \epsilon < 1$, finds an integer vector p and an integer q satisfying*

$$\|a - \frac{1}{q}p\| < \frac{\epsilon}{q} \quad \text{and} \quad 1 \leq q \leq 2^{\frac{1}{4}n(n+1)} \epsilon^{-n}. \tag{15}$$

This can be seen by applying the basis reduction method to the matrix

$$A := \begin{bmatrix} 1 & & & & \alpha_1 \\ & \ddots & & 0 & \vdots \\ & 0 & \ddots & & \vdots \\ & & & 1 & \alpha_n \\ \hline 0 & \cdots\cdots & & 0 & 2^{-\frac{1}{4}n(n+1)}\epsilon^{n+1} \end{bmatrix}, \qquad (16)$$

denoting $a =: (\alpha_1, \ldots, \alpha_n)^T$.

FRANK and TARDOS [6] showed that this simultaneous diophantine approximation method yields so-called strongly polynomial algorithms. The ellipsoid method discussed in Section 3 can derive a polynomial-time algorithm for the optimization problem (8) from a polynomial-time algorithm for the separation problem (10), and vice versa. The polynomial-time algorithms for (8) derived perform a number of arithmetic operations, which number is bounded by a polynomial in (9). (*Arithmetic operations* here are: addition, subtraction, multiplication, division and comparison of numbers.) Although this does not conflict the definition of 'polynomial-time', it would be preferable if the size of the 'cost' function c only influences the size of the numbers occurring when executing the algorithm, but not the number of arithmetic operations. Therefore, an algorithm for the optimization problem (8) is called *strongly polynomial* if it consists of a number of arithmetic operations, bounded by a polynomial in $|V|+|E|$, on numbers of size bounded by a polynomial in $|V|+|E|+\text{size}(c)$.

FRANK and TARDOS now showed however the equivalence of the two concepts when applied to (8):

THEOREM 3. *For any class ($\mathcal{F}_G|G$ graph), there exists a polynomial-time algorithm for the optimization problem (8), if and only if there exists a strongly polynomial algorithm for (8).*

PROOF. The 'if' part being trivial, we sketch a proof of the 'only if' part. Suppose (8) is polynomially solvable for a certain class ($\mathcal{F}_G|G$ graph). Let $G = (V, E)$ and $c \in \mathbb{Q}^E$ be given as input for (8). Determine vectors $c_1, c_2, \ldots$ successively as follows. $c_1 := c$. Suppose $c_1, \ldots, c_i$ has been found. If $c_i \neq 0$, let

$$v := 2^{-5n^2} \left\lfloor \frac{2^{5n^2}}{\|c_i\|_\infty} c_i \right\rfloor, \qquad (17)$$

where $n := |E|$, and where $\lfloor \ \rfloor$ denotes component-wise lower integer parts. By the method of Theorem 2 we can find $u_i \in \mathbb{Z}^n$ and $q_i \in \mathbb{Z}$ such that

$$\left\| v - \frac{1}{q_i} u_i \right\|_\infty < \frac{1}{q_i} \cdot \frac{1}{2n} \quad \text{and} \quad 1 \leq q_i \leq 2^{\frac{1}{4}n(n+1)}(2n)^n \qquad (18)$$

(taking $\epsilon := 1/2n$). Note that

$$\|u_i\|_\infty \leq q_i \leq 2^{n^2}(2n)^n, \tag{19}$$

since $\|v\|_\infty = 1$. Let

$$c_{i+1} := c_i - \frac{\|c_i\|_\infty}{q_i} u_i. \tag{20}$$

If $c_{i+1} = \mathbf{0}$, stop. Otherwise, repeat with i replaced by $i+1$.

Since c_{i+1} has at least one 0-component more than c_i has, as one easily derives from (17), (18) and (20), the algorithm stops after $k \leq n$ iterations. Let $c_1, \ldots, c_k$ be the sequence generated. Note that by (20),

$$c = \frac{\|c_1\|_\infty}{q_1} u_1 + \frac{\|c_2\|_\infty}{q_2} u_2 + \ldots + \frac{\|c_k\|_\infty}{q_k} u_k. \tag{21}$$

Now define:

$$\tilde{c} := 2^{5n^2(k-1)} u_1 + 2^{5n^2(k-2)} u_2 + \ldots + 2^{5n^2} u_{k-1} + u_k. \tag{22}$$

The vector $\tilde{c}$ has the following property:

$$\text{for each vector } x \in \{0, \pm 1\}^n: \text{ if } c^T x < 0 \text{ then } \tilde{c}^T x < 0. \tag{23}$$

Indeed, let $x \in \{0, \pm 1\}^n$ with $c^T x < 0$. Choose the smallest i with $u_i^T x \neq 0$ (i exists by (21)). Then

$$c_i^T x = (c - \frac{\|c_1\|_\infty}{q_1} u_1 - \ldots - \frac{\|c_{i-1}\|_\infty}{q_{i-1}} u_{i-1})^T x = c^T x < 0. \tag{24}$$

Hence

$$u_i^T x = (u_i - q_i v)^T x + q_i(v - c_i)^T x + q_i c_i^T x \tag{25}$$

$$< \|u_i - q_i v\|_\infty \cdot \|x\|_1 + q_i \cdot \|v - c_i\|_\infty \cdot \|x\|_1$$

$$\leq \frac{1}{2n} n + 2^{n^2}(2n)^n 2^{-5n^2} n \leq 1,$$

implying $u_i^T x \leq -1$ (as u_i is integral and $u_i^T x \neq 0$). Therefore,

$$\tilde{c}^T x = 2^{5n^2(k-i)} u_i^T x + 2^{5n^2(k-i-1)} u_{i+1}^T x + \ldots + u_k^T x \tag{26}$$

$$\leq -2^{5n^2(k-i)} + n \cdot 2^{5n^2(k-i-1)} \cdot 2^{4n^2}$$

$$= 2^{5n^2(k-i)}(-1 + n \cdot 2^{-n^2}) < 0$$

(using $u_j^T x \leq \|u_j\|_\infty \cdot \|x\|_1 \leq 2^{n^2}(2n)^n n \leq 2^{4n^2}$ - cf. (19)). This proves (23).

Having determined $\tilde{c}$, give the input $G = (V, E)$ and $\tilde{c} \in \mathbb{Z}^E$ to the polynomial-time algorithm for (8). It gives a set F in $\mathcal{F}_G$ maximizing $\Sigma_{e \in F} \tilde{c}(e)$. Then F also maximizes $\Sigma_{e \in F} c(e)$. For suppose $\Sigma_{e \in F'} c(e) > \Sigma_{e \in F} c(e)$ for some $F' \in \mathcal{F}_G$. Then $c^T(\chi^F - \chi^{F'}) < 0$. By (23), $\tilde{c}^T(\chi^F - \chi^{F'}) < 0$, contradicting the fact that F maximizes $\Sigma_{e \in F} \tilde{c}(e)$.

The whole procedure consists of a number of arithmetic operations bounded by a polynomial in $|V| + |E|$. Indeed, v in (17) can be determined by binary search by $5n^2 + 1$ comparisons (for each coordinate). The method of Theorem

2 applied to v and $\epsilon = 1/2n$ takes time bounded by a polynomial in size$(v) = \mathcal{O}(n^3)$ and size$(\epsilon) = \mathcal{O}(\log n)$. Finally, the algorithm for the optimization problem applied to G and $\tilde{c}$ takes time bounded by a polynomial in $|V|+|E|$ and size$(\tilde{c}) = \mathcal{O}(n^6)$ (by (19) and (21)). Concluding, we have a strongly polynomial algorithm for (8). □

A similar result holds for the separation problem (10).

5. Totally unimodular matrices and bipartite graphs

We now come to some concrete examples of polyhedral characterizations. A prime technique in deriving polyhedral results is based on 'total unimodularity' of matrices. A matrix is called *totally unimodular* if each subdeterminant belongs to $\{0, +1, -1\}$. In particular, each entry in a totally unimodular matrix belongs to $\{0, +1, -1\}$.

The following is not difficult to see.

THEOREM 4. *Let A be a totally unimodular $m \times n$-matrix, and let b be an integral column vector in $\mathbb{R}^m$. Then each vertex of the polyhedron $\{x | Ax \leq b\}$ is integral.*

PROOF. Let x^* be a vertex of $\{x | Ax \leq b\}$. Then there exists a nonsingular $m \times m$-submatrix A' of A, with corresponding part b' of b, so that $A'x^* = b'$. Hence $x^* = (A')^{-1}b'$. As $\det A' = \pm 1$, it follows that x^* is integral. □

This theorem and extensions characterizing total unimodularity were given by Hoffman and Kruskal [13].

Let $G = (V, E)$ be a *bipartite graph,* i.e., an undirected graph whose vertex set V can be split into two classes V' and V'' so that each edge consists of a vertex in V' and a vertex in V''. Let A be the *incidence matrix* of G, i.e., A is the $V \times E$- matrix with 1 in position (v, e) if $v \in e$, and 0 otherwise.

THEOREM 5. *The incidence matrix of a bipartite graph is totally unimodular.*

PROOF. Let B be an $m \times m$-submatrix of A. We show $\det B \in \{0, \pm 1\}$ by induction on m, the case $m = 1$ being trivial. If B contains an all-zero column, then $\det B = 0$. If B contains a column with exactly one 1, we can expand $\det B$ by this column, yielding $\det B = \pm \det B'$ for some $(m-1) \times (m-1)$-submatrix B' of B. Then by induction $\det B \in \{0, \pm 1\}$. If each column of B contains exactly two 1's, we can decompose B as $\begin{bmatrix} B' \\ B'' \end{bmatrix}$ so that each column of B' has exactly one 1, and similarly for B'' (possibly after permuting rows of B). Then $(1, \ldots, 1, -1, \ldots, -1) \begin{bmatrix} B' \\ B'' \end{bmatrix} = \mathbf{0}$, and hence $\det B = 0$. □

Theorems 4 and 5 have some direct consequences. For any graph $G = (V, E)$, the polytope conv.hull $\{\chi^M | M \text{ matching}\}$ is called the *matching polytope* of G.

THEOREM 6. *Let $G=(V,E)$ be a graph. Then the matching polytope of G is equal to the set of all vectors $x \in \mathbb{R}^E$ satisfying*

(i) $x_e \geq 0 \quad (e \in E)$,

(ii) $\sum_{e \ni v} x_e \leq 1 \quad (v \in V)$, (27)

if and only if G is bipartite.

PROOF. 'if': If G is bipartite, its incidence matrix A is totally unimodular, and hence also the matrix $\begin{bmatrix} -I \\ A \end{bmatrix}$ is totally unimodular. Since the system (27) is the same as $\begin{bmatrix} -I \\ A \end{bmatrix} x \leq \begin{bmatrix} \mathbf{0} \\ \mathbf{1} \end{bmatrix}$, we know that the polytope P defined by (27) has integral vertices only (Theorem 4). Since the integral vectors satisfying (27) are exactly the vectors χ^M for matchings M, we know that P is equal to the matching polytope of G.

'only if': If G is not bipartite, it has an odd circuit C. Let $x \in \mathbb{R}^E$ be defined by $x_e = \frac{1}{2}$ if e belongs to C, and $x_e = 0$ otherwise. Then x satisfies (27), but x does not belong to the matching polytope of G, as one easily checks. □

This theorem immediately yields a strongly polynomial algorithm for finding a maximum-weighted matching in a bipartite graph $G=(V,E)$ (which problem is one of the variants of the *optimal assignment problem*): given a weight function $c \in \mathbb{Z}^E$, a matching M in G maximizing $\Sigma_{e \in M} c(e)$ can be found by solving the linear program of maximizing $c^T x$ over (27).

This can be derived also from Theorems 1 and 3, since solving the separation problem for matching polytopes of bipartite graphs just means testing if a given vector x satisfies (27); these constraints can be checked one by one in polynomial time (there are $|V|+|E|$ constraints). This does not reflect the full power of Theorem 1 - we shall see a better illustration in the next section.

A similar result holds for the *coclique polytope* of a graph G, being conv.hull $\{\chi^C | C \text{ coclique}\}$.

THEOREM 7. *Let $G=(V,E)$ be a graph. Then the coclique polytope of G is equal to the set of all vectors $x \in \mathbb{R}^E$ satisfying*

(i) $x_v \geq 0 \quad (v \in V)$,

(ii) $\sum_{v \in e} x_v \leq 1 \quad (e \in E)$, (28)

if and only if G is bipartite.

PROOF. Similarly to the proof of Theorem 6 (note that clearly also the transpose A^T of the incidence matrix of a bipartite graph is totally unimodular).
□

Again, one can derive from this that for bipartite graphs a maximum weighted coclique can be found in polynomial time.

The following two related results are classical. The *perfect matching polytope* of a graph is the polytope conv.hull$\{\chi^M | M \text{ perfect matching}\}$. A *perfect matching* is a matching covering all vertices of the graph exactly once.

THEOREM 8 (BIRKHOFF-VON NEUMANN THEOREM). *Let $G = (V, E)$ be a bipartite graph. Then the perfect matching polytope is equal to the set of all vectors x in $\mathbb{R}^E$ satisfying*

(i) $x_e \geq 0$ $(e \in E)$,

(ii) $\sum_{e \ni v} x_e = 1$ $(v \in V)$.   (29)

PROOF. The theorem follows from the total unimodularity of the matrix

$$\begin{bmatrix} -I \\ A \\ -A \end{bmatrix}, \quad (30)$$

where A is the incidence matrix of G. □

This theorem is better known in the following equivalent formulation: each doubly stochastic matrix is a convex combination of permutation matrices. (A matrix is *doubly stochastic* if it is nonnegative and if each row sum and each column sum is equal to 1.)

Theorem 8 yields the polynomial-time solvability of the problem of finding a maximum (and similarly, a minimum) weighted perfect matching in a bipartite graph.

THEOREM 9 (KÖNIG-EGERVÁRY THEOREM). *Let $G = (V, E)$ be a bipartite graph. Then*

(i) $\max\{|M| \, | M \text{ matching}\} = \min\{|W| \, | W \subseteq V; \forall e \in E : \exists v \in W : v \in e\}$;   (31)

(ii) $\max\{|C| \, | C \text{ coclique}\} = \min\{|F| \, | F \subseteq E; \forall v \in V : \exists e \in F : v \in e\}$.

PROOF. Let A be the incidence matrix of G. Then by the total unimodularity of A:

$$\max\{|M| \, | M \text{ matching}\} = \max\{\mathbf{1}^T x | x \geq 0; Ax \leq \mathbf{1}; x \text{ integral}\}$$

$$= \max\{\mathbf{1}^T x | x \geq 0; Ax \leq \mathbf{1}\} = \min\{y^T \mathbf{1} | y \geq 0; y^T A \geq \mathbf{1}^T\}$$

$$= \min\{y^T \mathbf{1} | y \geq 0; y^T A \geq \mathbf{1}^T; y \text{ integral}\} \quad (32)$$

$$= \min\{|W| \mid W \subseteq V; \forall e \in E: \exists v \in W: v \in e\}.$$

This shows (i). Equation (ii) is shown similarly. □

REMARK. Similar results can be derived for flows in directed graphs (like the max-flow min-cut theorem), using the fact that any $\{0, +1, -1\}$-matrix with in each column at most one 1 and at most one -1, is totally unimodular.

Here we mention SEYMOUR'S deep result [24] that each totally unimodular matrix can be decomposed, in a certain way, into matrices described in the previous sentence and into two certain totally unimodular 5×5-matrices. It yields a polynomial-time test for the total unimodularity of matrices (clearly, checking all subdeterminants would require exponential time).

6. THE MATCHING POLYTOPE OF AN ARBITRARY GRAPH

If G is not bipartite, the inequalities (27) are not enough to determine the matching polytope. A famous theorem of EDMONDS [5] gives the inequalities determining the matching polytope of a not-necessarily bipartite graph. Similarly, Edmonds characterized the perfect matching polytope, which we discuss first. We follow the proof of [21].

THEOREM 10 (EDMONDS' MATCHING POLYTOPE THEOREM). *For any graph $G = (V,E)$ the perfect matching polytope is equal to the set of vectors x in $\mathbb{R}^E$ satisfying*

$$\begin{aligned}&\text{(i)} \quad x_e \geq 0 \quad (e \in E),\\&\text{(ii)} \quad \sum_{e \ni v} x_e = 1 \quad (v \in V), \quad (33)\\&\text{(iii)} \quad x(\delta(U)) \geq 1 \quad (U \subseteq V, |U| \text{ odd}).\end{aligned}$$

Here $\delta(U)$ denotes the set of edges e in E with $|e \cap U| = 1$, and $x(\delta(U)) := \sum_{e \in \delta(U)} x_e$.

PROOF. Let P be the perfect matching polytope of G, and let Q be the set of vectors satisfying (33). As $\chi^M \in Q$ for each perfect matching M, it follows that $P \subseteq Q$ - the content of the theorem is the converse inclusion.

Let G be a smallest graph with $Q \not\subseteq P$ (that is, with $|V| + |E|$ minimal), and let x be a vertex of Q not contained in P. Then $0 < x_e < 1$ for all e in E - otherwise we could delete e from G to obtain a smaller counterexample. Moreover, $|E| > |V|$ - otherwise, either G is disconnected (in which case one of the components of G will be a smaller counterexample), or G has a point v of degree one (in which case the edge e incident to v has $x_e = 1$), or G is an even circuit (for which the theorem trivially holds).

Since x is a vertex of Q, there are $|E|$ independent constraints among (33) satisfied by x with equality, and hence there exists a $U \subseteq V$ with $|U|$ odd, $|U| \geq 3, |V \setminus U| \geq 3$ and $x(\delta(U)) = 1$. Let G_1 and G_2 arise from G by contracting U and $V \setminus U$, respectively, and let x_1 and x_2 be the corresponding projections of x onto the edge sets of G_1 and G_2, respectively. Since x_1 and x_2 satisfy inequalities (33) for the smaller graphs G_1 and G_2, respectively, it follows that x_1 and x_2 can be decomposed as convex combinations of characteristic vectors of perfect matchings in G_1 and G_2, respectively. These decompositions can be easily glued together to form a decomposition of x as a convex combination of perfect matchings, contradicting our assumption.

(This glueing can be done, e.g., as follows. By the rationality of x (as it is a vertex of Q), there exists a natural number K such that, for $i=1,2$, Kx_i is the sum of the characteristic vectors of the perfect matchings $M_1^i, \ldots, M_K^i$ of G_i (possibly with repetitions). Since, for each e in $\delta(U)$, e is contained in $Kx(e)$ of the M_j^1 as well as in $Kx(e)$ of the M_j^2, we may assume that $M_j^1 \cap M_j^2 \neq \emptyset$ for $j=1,\ldots,K$. It follows that Kx is the sum of the characteristic vectors of the perfect matchings $M_1^1 \cup M_1^2, \ldots, M_K^1 \cup M_K^2$ of G, and hence that x itself is a convex combination of perfect matchings in G.) □

Application of Theorem 1 now becomes more illustrative than in Section 5. By Theorem 1, we can find a maximum weighted perfect matching in a graph in polynomial time, if we can solve the separation problem for the perfect matching polytope in polynomial time. This last can be shown as follows (following PADBERG and RAO [18]). For a given $x \in \mathbf{Q}^E$ we have to test if x satisfies (33). Testing the inequalities in (i) and (ii) can be done easily by checking them one by one. If one of them is not satisfied, we know that x does not belong to the perfect matching polytope, and the violated constraint gives a separating hyperplane. So we may assume that x satisfies (33) (i) and (ii). If $|V|$ is odd, then clearly (33) (iii) is not satisfied for $U:=V$. So we may assume that $|V|$ is even. We cannot check the constraints in (iii) one by one in polynomial time, simply because there are exponentially many of them. Yet, there is a polynomial-time method of checking them. Indeed, first note that from Ford-Fulkerson's max-flow min-cut algorithm we can derive a polynomial-time algorithm having the following as in- and output:

> *input*: subset W of V;
> *output*: a subset T of V such that $W \cap T \neq \emptyset \neq W \setminus T$ and such that (34) $x(\delta(T))$ is as small as possible.

To see this, consider x as a capacity function on E, and determine for each pair $r,s \in W$, a cut of minimum capacity separating r and s. That is, we find a subset $T_{r,s}$ of V so that $r \in T_{r,s}$, $s \notin T_{r,s}$ and such that $\text{cap}(\delta(T_{r,s})):=x(\delta(T_{r,s}))$ is minimal. Taking $T:=T_{r,s}$ for that pair r,s for which $\text{cap}(\delta(T_{r,s}))$ is as small as possible, we obtain T as required.

We next describe recursively an algorithm with the following in- and output:

> *input*: subset W of V with $|W|$ even;
> *output*: subset U of V such that $|W \cap U|$ is odd and such that $x(\delta(U))$ (35) is as small as possible.

First we find with the algorithm (34) a subset T of V with $W \cap T \neq \emptyset \neq W \setminus T$ and such that $x(\delta(T))$ is minimal. If $|W \cap T|$ is odd we are done. If $|W \cap T|$ is even, call, recursively, the algorithm (35) for the inputs $W \cap T$ and $W \cap \overline{T}$, respectively, where $\overline{T} := V \setminus T$. Let it yield a subset U' of V such that $|W \cap T \cap U'|$ is odd and $x(\delta(U'))$ is minimal, and a subset U'' of V such that $|W \cap \overline{T} \cap U''|$ is odd and $x(\delta(U''))$ is minimal. Without loss of generality, $W \setminus T \not\subseteq U'$ (otherwise replace U' by $V \setminus U'$) and $W \setminus \overline{T} \not\subseteq U''$ (otherwise replace U'' by $V \setminus U''$).

We claim that if $x(\delta(T \cap U')) \leq x(\delta(\overline{T} \cap U''))$ then $U := T \cap U'$ is output of (35) for input W, and otherwise $U := \overline{T} \cap U''$ is output of (35) for input W. To see that this output is justified, suppose to the contrary that there exists a subset Y of V such that $|W \cap Y|$ is odd and $x(\delta(Y)) < x(\delta(T \cap U'))$ and $x(\delta(Y)) < x(\delta(\overline{T} \cap U''))$. Then either $|W \cap Y \cap T|$ is odd or $|W \cap Y \cap \overline{T}|$ is odd (since $|W \cap T|$ is even). *Case 1*: $|W \cap Y \cap T|$ is odd. Then $x(\delta(Y)) \geq x(\delta(U'))$, since U' is output of (35) for input $W \cap T$. Moreover, $x(\delta(T \cup U')) \geq x(\delta(T))$, since T is output of (34) for input W, and since $W \cap (T \cup U') \neq \emptyset \neq W \setminus (T \cup U')$. Therefore, we have a contradiction:

$$x(\delta(Y)) \geq x(\delta(U')) \geq x(\delta(T \cap U')) + x(\delta(T \cup U')) - x(\delta(T)) \qquad (36)$$
$$\geq x(\delta(T \cap U')) > x(\delta(Y))$$

(the second inequality follows since $x(\delta(A)) + x(\delta(B)) \geq x(\delta(A \cap B)) + x(\delta(A \cup B))$ for all subsets A and B of V). *Case 2*: $|W \cap Y \cap \overline{T}|$ is odd. Similarly.

Given the polynomiality of the algorithm for (34), it is not difficult to see that also the described algorithm for (35) has polynomially bounded running time.

As a consequence, we can test the inequalities (33) (iii) in polynomial time, which implies the polynomial-time solvability of the problem of finding a maximum weighted perfect matching. In fact, EDMONDS [5] gave a direct polynomial-time algorithm for this problem, yielding Theorem 10 as a by-product. We have followed the above line to illustrate the use of Theorem 1.

By a standard construction, Edmonds' characterization of the matching polytope can be derived from Theorem 10.

THEOREM 11. *For any graph $G = (V, E)$, the matching polytope is equal to the set of all vectors x in $\mathbb{R}^E$ satisfying*

$$\text{(i)} \quad x_e \geq 0 \qquad (e \in E),$$
$$\text{(ii)} \quad \sum_{e \ni v} x_e \leq 1 \qquad (v \in V), \qquad (37)$$
$$\text{(iii)} \quad \sum_{e \subseteq U} x_e \leq \lfloor \tfrac{1}{2} |U| \rfloor \qquad (U \subseteq V, |U| \text{ odd}).$$

PROOF. Again it is clear that each vector in the matching polytope satisfies (37), as χ^M satisfies (37) for each matching M. To see that the inequalities (37) are enough, let $x \in \mathbb{R}^E$ satisfy (37). Let $G^* = (V^*, E^*)$ be a disjoint copy of G, where the copy of vertex v will be denoted by v^*, and the copy of edge

$e = \{v, w\}$ will be denoted by $e^* = \{v^*, w^*\}$. Let $\tilde{G}$ be the graph with vertex set $V \cup V^*$ and with edge set $E \cup E^* \cup \{\{v, v^*\} | v \in V\}$. Define $\tilde{x}(e) := \tilde{x}(e^*) := x(e)$ for e in E, and $\tilde{x}(\{v, v^*\}) := 1 - x(\delta(v))$ for v in V. Now conditions (33) are easily derived for $\tilde{x}$ with respect to $\tilde{G}$. Constraints (i) and (ii) are trivial. To prove (iii) in (33), we have to show, for $V_1, V_2 \subseteq V$ with $|V_1| + |V_2|$ odd, that $\tilde{x}(\delta(V_1 \cup V_2^*)) \geq 1$. Indeed, we may assume, without loss of generality, that $|V_1 \setminus V_2|$ is odd. Hence

$$\tilde{x}(\delta(V_1 \cup V_2^*)) = \tilde{x}(\delta(V_1 \setminus V_2)) + \tilde{x}(\delta(V_2^* \setminus V_1^*)) \geq \tilde{x}(\delta(V_1 \setminus V_2))$$
$$= |V_1 \setminus V_2| - 2 \cdot \sum_{e \subseteq V_1 \setminus V_2} x_e \geq 1, \qquad (38)$$

by (37) (iii).

Hence $\tilde{x}$ is a convex combination of perfect matchings in $\tilde{G}$. By restriction to x and G it follows that x is a convex combination of matchings in G. □

In a way similar to above one can derive a polynomial-time algorithm finding a maximum weighted matching.

Related to Theorem 11 is the following min-max relation due to TUTTE [26] and BERGE [1].

THEOREM 12 (TUTTE-BERGE FORMULA). *For any graph* $G = (V, E)$

$$\max\{|M| \,|\, M \text{ matching}\} = \min_{U \subseteq V} \frac{|V| + |U| - \mathcal{O}(V \setminus U)}{2} \qquad (39)$$

where $\mathcal{O}(V \setminus U)$ *denotes the number of odd components of the graph induced by* $V \setminus U$.

The minimum here can be easily seen to be equal to:

$$\min\{|U| + \sum_{i=1}^{t} \lfloor \tfrac{1}{2}|V_i| \rfloor \,|\, U, V_1, \ldots, V_t \subseteq V, \text{ so that each edge} \qquad (40)$$
intersects U or is contained in one of the V_i\}.

The content of the Tutte-Berge formula is that when we write $\max\{|M| \,|\, M$ matching$\}$ equivalently as maximizing $\mathbf{1}^T x$ over (37), we obtain a linear program with integral optimal primal *and* dual solutions.

7. CUTTING PLANES

Quite often the problem of characterizing the convex hull of certain $\{0, 1\}$-vectors amounts to characterizing, for some polytope P, the polytope

$$P_I := \text{conv.hull } \{x \in P \,|\, x \text{ integral}\}. \qquad (41)$$

P_I is called the *integer hull* of P. E.g., if $G = (V, E)$ is a graph, and

$$P := \{x \in \mathbf{R}^E \,|\, x_e \geq 0 \ (e \in E), \sum_{e \ni v} x_e \leq 1 \ (v \in V)\}, \qquad (42)$$

the integral vectors in P are exactly the characteristic vectors of matchings, and hence P_I is equal to the matching polytope of G. Similarly, for

$$P := \{x \in \mathbf{R}^V | x_v \geq 0 \ (v \in V), \sum_{v \in e} x_v \leq 1 \ (e \in E)\}, \tag{43}$$

P_I is the coclique polytope of G.

There is a way of deriving the inequalities determining P_I from those determining P - the *cutting plane method.* Its basics were given by GOMORY [10]. The following description is due to CHVÁTAL [2] and SCHRIJVER [20].

Clearly, if H is a *rational halfspace,* i.e., H is of form

$$H = \{x \in \mathbf{R}^n | a^T x \leq \beta\}, \tag{44}$$

where $a \in \mathbf{Q}^n$, $a \neq 0$, $\beta \in \mathbf{Q}$, we may assume without loss of generality that a is integral, and that the components of a are relatively prime. In that case:

$$H_I = \{x \in \mathbf{R}^n | a^T x \leq \lfloor \beta \rfloor\}. \tag{45}$$

H_I arises from H by shifting its bounding hyperplane until it contains integral vectors.

Now define for any set P:

$$P' := \bigcap_{H \supseteq P} H_I, \tag{46}$$

where H ranges over all rational halfspaces containing P. Since $H \supseteq P$ implies $H_I \supseteq P_I$, it follows that $P_I \subseteq P'$. It can be shown that if P is a rational polyhedron (i.e., a polyhedron determined by linear inequalities with rational coefficients), then P' is a polyhedron again.

To P' we can apply this operation again, yielding P''. Generally $P'' \neq P'$ - consider e.g. the following example.

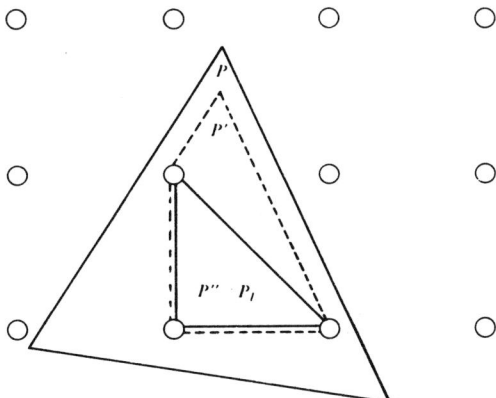

FIGURE 1

So there is a sequence of polyhedra containing P_I:

$$P \supseteq P' \supseteq P'' \supseteq P''' \supseteq \ldots \supseteq P_I. \tag{47}$$

Denoting the $(t+1)$-th set in this sequence by $P^{(t)}$, the following can be shown.

THEOREM 13. *For each rational polyhedron P there exists a number t such that $P^{(t)} = P_I$.*

The theorem is the theoretical essence of the cutting plane method of Gomory. The equation $ax = \lfloor \beta \rfloor$ defining H_I, or more strictly the hyperplane $\{x | ax = \lfloor \beta \rfloor\}$, is called a *cutting plane*.

The smallest t for which $P^{(t)} = P_I$ can be considered as a measure for the complexity of P_I relative to that of P. In a sense, P' is conceptually near to P, P'' to P', etc.

Let us study some specific polyhedra related to graphs. Let $G = (V, E)$ be an undirected graph, and let $P \subseteq \mathbb{R}^E$ be the polytope determined by the inequalities

$$\begin{aligned}&\text{(i)} \quad x_e \geq 0 \quad (e \in E), \\ &\text{(ii)} \quad \sum_{e \ni v} x_e \leq 1 \quad (v \in V).\end{aligned} \qquad (48)$$

So P_I is the matching polytope of G. By Theorem 6, $P = P_I$ if and only if G is bipartite. It is not difficult to see that for each graph G, P' is the set of all vectors x satisfying (48) and satisfying

$$\sum_{e \subseteq U} x_e \leq \lfloor \tfrac{1}{2}|U| \rfloor \quad (U \subseteq V, |U| \text{ odd}). \qquad (49)$$

(Of course, there are infinitely many halfspaces H containing P, but the corresponding inequalities $ax \leq \lfloor \beta \rfloor$ all are implied by the inequalities in (48) and (49).) So Theorem 11 in fact tells us that $P' = P_I$ for each graph G.

Next consider for any graph $G = (V, E)$ the polytope $P \subseteq \mathbb{R}^V$ determined by the inequalities:

$$\begin{aligned}&\text{(i)} \quad x_v \geq 0 \quad (v \in V), \\ &\text{(ii)} \quad \sum_{v \in e} x_v \leq 1 \quad (e \in E).\end{aligned} \qquad (50)$$

For this P, P_I is the coclique polytope of G. By Theorem 7, $P = P_I$ if and only if G is bipartite. It is not difficult to check that for any graph G, the polytope P' is the set of vectors x satisfying (50) and satisfying

$$\sum_{v \in V(C)} x_v \leq \lfloor \tfrac{1}{2}|V(C)| \rfloor \quad (C \text{ odd circuit}), \qquad (51)$$

where $V(C)$ is the vertex set of C, and where an odd circuit is a circuit C with $|V(C)|$ odd.

CHVÁTAL [2] has shown that there exists no fixed t such that $P^{(t)} = P_I$ for each graph G. The problem of finding a largest coclique in a graph is *NP*-complete, and hence probably not polynomially solvable. Therefore, by Theorem 1, probably there is no 'decent' description of the coclique polytope for all graphs. It is conjectured that for each fixed t, when we restrict ourselves to graphs which have $P^{(t)}=P_I$, the problem of finding a maximum weighted coclique is polynomially solvable (in fact, this problem can be shown to belong to $NP \cap$ co-NP). The conjecture is true for $t=0$ and $t=1$ (using Theorem 1). If we want to show it for $t=2$, by Theorem 1 it suffices to show that the following problem is polynomially solvable: decide if a given vector $x \in \mathbb{R}^V$ belongs to P'', and if not, find a separating hyperplane.

In Section 9 we shall see a class of graphs with $P'=P_I$. As a preparation, we discuss in Section 8 another geometric tool.

8. BINARY CONFIGURATIONS

We now come to a geometric method of a nature different from those discussed above. Let us call a set $x_1, \ldots, x_k$ of vectors in some space $GF(2)^n$ a *binary configuration*. Usually, the zero-vector will not be among $x_1, \ldots, x_k$, and hence we can consider a binary configuration as a configuration in $PG(d, 2)$, the d-dimensional projective space over $GF(2)$.

A well-known binary configuration is the *Fano-configuration* ($=PG(2,2)$) which is the binary configuration represented by the columns of

$$\begin{bmatrix} 1 & 0 & 0 & 1 & 1 & 0 & 1 \\ 0 & 1 & 0 & 1 & 0 & 1 & 1 \\ 0 & 0 & 1 & 0 & 1 & 1 & 1 \end{bmatrix} \tag{52}$$

and whose 7 points and 7 lines can be represented as:

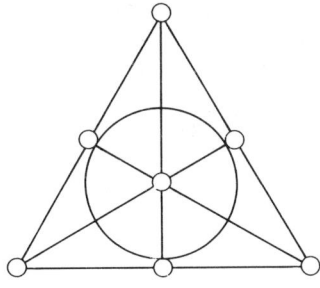

FIGURE 2

The lines are represented by 6 line segments and one circle. The reader familiar with the Fano-configuration might have tried to draw it on the paper in such a way that *all* 7 lines become straight line segments, so that not all 7 points are on one and the same line in the plane. After some trials one will be convinced that this is not possible, and it is not hard to show this algebraically.

In fact, Fano is in a sense a critical example. A famous and deep theorem of TUTTE [26] states that a binary configuration can be embedded in euclidean space so that each subset of points span a space of the same dimension in the binary space as they do in euclidean space, if and only if the binary configuration does not 'contain' the Fano-configuration or its 'dual'.

We shall make terms more precise. Call a binary configuration $x_1, \ldots, x_k$

in $GF(2)^n$ *embeddable in euclidean space* if there exists a function $\phi:\{x_1,\ldots,x_k\}\to\mathbb{R}^n$ so that for each subset T of $\{x_1,\ldots,x_k\}$, the dimension of $<T>$ in $GF(2)^n$ is equal to the dimension of $<\phi[T]>$ in $\mathbb{R}^n$. The function ϕ is called an *embedding*.

Deleting x_1 from $x_1,\ldots,x_k$ means replacing $x_1,\ldots,x_k$ by $x_2,\ldots,x_k$. *Projecting along* x_1 or *contracting* x_1 means replacing $x_1,\ldots,x_k$ by

$$x_2/<x_1>,\ldots,x_k/<x_1>, \tag{53}$$

where $../<x_1>$ means projecting $..$ into the quotient space $GF(2)^n/<x_1>$. Two binary configurations $x_1,\ldots,x_k$ and $x'_1,\ldots,x'_k$ are called *geometrically the same* if there is a linear transformation bringing $x_1,\ldots,x_k$ one-to-one to $x'_1,\ldots,x'_k$. Thus the Fano-configuration is geometrically the same as the set of columns of

$$\begin{bmatrix} 1 & 1 & 1 & 0 & 0 & 0 & 1 \\ 1 & 0 & 0 & 1 & 1 & 0 & 1 \\ 0 & 1 & 0 & 1 & 0 & 1 & 1 \\ 0 & 0 & 1 & 0 & 1 & 1 & 1 \end{bmatrix}. \tag{54}$$

A binary configuration Y is called a *minor* of a binary configuration X, if Y can be obtained from X by deletion, projection and permutation of vectors, up to being geometrically the same.

Trivially, embeddability in euclidean space is maintained under deletion. It is also not difficult to see that it is maintained under projection. Indeed, if $\phi:\{x_1,\ldots,x_k\}\to\mathbb{R}^n$ forms an embedding, then also $x_i/<x_1>\mapsto\phi(x_i)/<\phi(x_1)>$ forms an embedding. It follows that embeddability in euclidean space is maintained under taking minors.

The *dual* of the Fano-configuration is the configuration represented by the columns of

$$\begin{bmatrix} 1 & 0 & 0 & 0 & 1 & 1 & 0 \\ 0 & 1 & 0 & 0 & 1 & 0 & 1 \\ 0 & 0 & 1 & 0 & 0 & 1 & 1 \\ 0 & 0 & 0 & 1 & 1 & 1 & 1 \end{bmatrix}. \tag{55}$$

Geometrically, the dual of the Fano-configuration is formed by the 7 points obtained from $PG(3,2)$ by deleting one projective plane and one further (arbitrary) point. Also this configuration is not embeddable in euclidean space.

Now Tutte's theorem is:

THEOREM 14. *A binary configuration* $x_1,\ldots,x_k$ *is embeddable in euclidean space, if and only if it has no minor equal to the Fano-configuration or its dual.*

In order to interprete and use this difficult theorem, we first make a further study of binary configurations. To each binary configuration $x_1,\ldots,x_k$ we can associate the binary space or *binary code* C of all vectors z in $GF(2)^k$

satisfying $[x_1, \ldots, x_k]z = \mathbf{0}$ (considering $x_1, \ldots, x_k$ as column vectors). Clearly, each linear subspace of $GF(2)^k$ is associated in this way to some binary configuration. Two binary configurations are geometrically the same if and only if the associated binary codes are the same.

The binary configuration $y_1, \ldots, y_k$ is called *dual* to $x_1, \ldots, x_k$ if the associated binary codes are each others orthogonal complements. Note that the well-known Hamming code is associated this way with the Fano-configuration, and the dual Hamming code to the dual of the Fano-configuration.

If C is the binary code associated to the binary configuration $x_1, \ldots, x_k$, and if we delete x_1, the associated binary code becomes $\{z \mid \begin{bmatrix} 0 \\ z \end{bmatrix} \in C\}$; if we would project along x_1, the associated code becomes $\{z \mid \begin{bmatrix} 0 \\ z \end{bmatrix} \in C \text{ or } \begin{bmatrix} 1 \\ z \end{bmatrix} \in C\}$.

Thus a binary configuration contains the Fano-configuration or its dual as a minor, if and only if by these operations the associated binary code can be transformed into the Hamming code or its dual.

This is all quite standard linear algebra. More specific is the following definition. A subspace C of $GF(2)^n$ is said to be *orientable* if we can associate with each $x \in C$ a vector x' in $\{0, \pm 1\}^n$ and with each $y \in C^\perp$ a vector y'' in $\{0, \pm 1\}^n$ in such a way that:

(i) $\forall x \in C: x$ and x' have the same support;
(ii) $\forall y \in C: y$ and y'' have the same support; (56)
(iii) $\forall x \in C, \forall y \in C^\perp : (x')^T y'' = 0$.

The following theorem now is not so difficult to prove:

THEOREM 15. *A binary configuration is embeddable in euclidean space, if and only if the associated binary code is orientable.*

REMARK. Another deep theorem characterizing binary configurations embeddable in euclidean space is due to SEYMOUR [24]. To describe this we need some concepts.

A binary configuration is called *graphic* if it is geometrically the same as a binary configuration $x_1, \ldots, x_k$ where each vector x_i has exactly two 1's. It follows that the associated binary code is the 'cycle space' of a graph. A binary configuration is *cographic* if it is the dual of a graphic configuration. So the associated binary code is the 'cocycle space' of a graph. It is not difficult to see that graphic and cographic configurations are embeddable in euclidean space.

Let be given two binary configurations $x_1, \ldots, x_k$ and $y_1, \ldots, y_t$, not containing the zero-vector, and let $d := \dim \langle x_1, \ldots, x_k \rangle + \dim \langle y_1, \ldots, y_t \rangle$ (where $\langle .. \rangle$ denotes projective space generated by .., and dim denotes projective dimension).

First, the two configurations can be embedded into $PG(d, 2)$ so that $\langle x_1, \ldots, x_k \rangle \cap \langle y_1, \ldots, y_t \rangle = \emptyset$. If $k \geq 1$, $t \geq 1$, then the binary configuration $x_1, \ldots, x_k, y_1, \ldots, y_t$ is called the *1-sum* of $x_1, \ldots, x_k$ and

$y_1, \ldots, y_t$.

Second, the two configurations can be embedded into $PG(d, 2)$ so that $x_1 = y_1$ and $\langle x_1, \ldots, x_k \rangle \cap \langle y_1, \ldots, y_t \rangle = \{x_1\}$. If $k \geq 3$, $t \geq 3$, then the binary configuration $x_2, \ldots, x_k, y_2, \ldots, y_k$ is called a *2-sum* of $x_1, \ldots, x_k$ and $y_1, \ldots, y_t$.

Third, let x_1, x_2, x_3 form a line and let y_1, y_2, y_3 form a line. Then the two configurations can be embedded into $PG(d, 2)$ so that $x_1 = y_1$, $x_2 = y_2$, $x_3 = y_3$ and $\langle x_1, \ldots, x_k \rangle \cap \langle y_1, \ldots, y_t \rangle = \{x_1, x_2, x_3\}$. If $k \geq 7$, $t \geq 7$, then the binary configuration $x_4, \ldots, x_k, y_4, \ldots, y_t$ is called a *3-sum* of $x_1, \ldots, x_k$ and $y_1, \ldots, y_t$.

Now Seymour's theorem is:

THEOREM 16. *A binary configuration is embeddable in euclidean space if and only if it can be obtained by making 1-, 2- and 3-sums from graphic configurations, cographic configurations, and the binary configuration made by the columns of*

$$\begin{bmatrix} 1 & 0 & 0 & 0 & 0 & 1 & 1 & 1 & 0 & 0 \\ 0 & 1 & 0 & 0 & 0 & 0 & 1 & 1 & 1 & 0 \\ 0 & 0 & 1 & 0 & 0 & 0 & 0 & 1 & 1 & 1 \\ 0 & 0 & 0 & 1 & 0 & 1 & 0 & 0 & 1 & 1 \\ 0 & 0 & 0 & 0 & 1 & 1 & 1 & 0 & 0 & 1 \end{bmatrix}. \tag{57}$$

Note that Tutte's theorem makes 'not embeddable in euclidean space' constructible, while Seymour's theorem makes 'embeddable in euclidean space' constructible.

Seymour's theorem has the following implication for totally unimodular matrices. Let A be a totally unimodular matrix. Then the binary configuration represented by the columns of the matrix $[I\ A]$ (forgetting the $-$ signs) is embeddable in euclidean space (as can be seen by not forgetting the $-$ signs). Hence Seymour's theorem implies that A can be decomposed into 'network matrices', their transposes, and the following two matrices:

$$\begin{bmatrix} -1 & 1 & -1 & 0 & 0 \\ 0 & -1 & 1 & -1 & 0 \\ 0 & 0 & -1 & 1 & -1 \\ -1 & 0 & 0 & -1 & 1 \\ 1 & -1 & 0 & 0 & -1 \end{bmatrix} \text{ and } \begin{bmatrix} 1 & 1 & 1 & 1 & 1 \\ 1 & 1 & 1 & 0 & 0 \\ 1 & 0 & 1 & 1 & 0 \\ 1 & 0 & 0 & 1 & 1 \\ 1 & 1 & 0 & 0 & 1 \end{bmatrix}. \tag{58}$$

The meaning of 'can be decomposed into' becomes clear after elaborating the meaning of 1-, 2-, and 3-sum. Seymour's theorem also yields a polynomial-time test for total unimodularity.

9. BINARY CONFIGURATIONS AND COMBINATORIAL OPTIMIZATION

There are several applications of the geometric results discussed in Section 8, e.g., to 2-commodity flows, the maximum cut problem, the Chinese postman problem, matchings - see SEYMOUR [23, 25]. We here restrict ourselves to one application to the coclique problem.

Let $G=(V,E)$ be an undirected graph, and consider the linear subspace C_G of $\{0,1\}\times\{0,1\}^E$ of all vectors $\begin{bmatrix}\epsilon\\ \chi^F\end{bmatrix}$ where F is a collection of edges so that each vertex of G is incident to an even number of edges in F, where χ^F denotes the characteristic vector of F in $\{0,1\}^E$, and where $\epsilon=0$ if $|F|$ is even and $\epsilon=1$ if $|F|$ is odd. Let K_G be the binary configuration (unique up to being geometrically the same) associated with C_G. Note that

$$C_G^\perp = \{\begin{bmatrix}\epsilon\\ \chi^F\end{bmatrix} | \epsilon = 0 \text{ and } F = \delta(W) \text{ for some } W \subseteq V, \text{ or } \epsilon = 1 \quad (59)$$
and $F = V\setminus\delta(W)$ for some $W\subseteq V\}$.

The following lemma is easy to check:

LEMMA. K_G does not contain the Fano-configuration or its dual as a minor, if and only if G has no subgraph isomorphic to one of the following graphs:

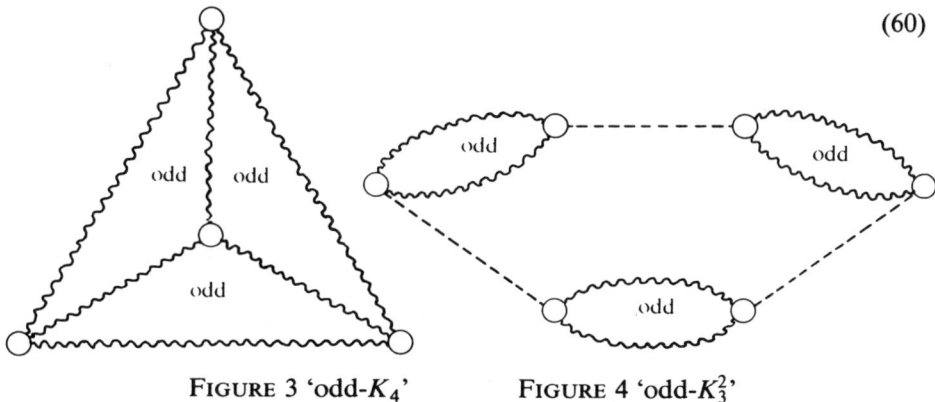

(60)

FIGURE 3 'odd-K_4'   FIGURE 4 'odd-K_3^2'

Here wriggled lines represent paths of positive length and dotted lines represent lines of positive or zero length; *odd* in a face means that the circuit enclosing it has an odd number of edges.

The following theorem is due to GERARDS [7].

THEOREM 17. *A graph G has no subgraph isomorphic to one of the graphs in (60), if and only if we can orient the edges of G in such a way that in each circuit, the number of edges directed one way differs by at most one from the number of edges directed in the other way.*

PROOF. The 'if' part follows easily, since the graphs in (60) do not have the required orientation, as one easily checks.

To see the 'only if' part, we may assume that G is connected. From the Lemma we know that K_G does not contain the Fano-configuration or its dual as a minor. By Tutte's theorem (Theorem 14), K_G is embeddable in euclidean space. Hence, by Theorem 15, C_G is orientable. Let the oriented vectors be indicated by $'$ and $''$ as in (56). So for each $x \in C_G$, $x' \in \{0, \pm 1\} \times \{0, \pm 1\}^E$ and for each $y \in C_G^\perp$, $y'' \in \{0, \pm 1\} \times \{0, \pm 1\}^E$. Without loss of generality

$$\begin{bmatrix} 1 \\ x^E \end{bmatrix}'' = \begin{bmatrix} 1 \\ x^E \end{bmatrix} \tag{61}$$

since we can multiply a certain coordinate by -1 throughout in all x' and all y'', not violating (56).

Let M be the matrix with columns all vectors

$$\begin{bmatrix} 0 \\ x^{\delta(v)} \end{bmatrix}'' \tag{62}$$

for v in V. Let T be a spanning tree in G. Since replacing y'' by $-y''$ does not change (56), we may assume that in any row of M corresponding to an edge in T there is exactly one 1 and one -1.

Now consider any other row, corresponding to edge $e \notin T$. There exist edges $e_1, \ldots, e_k$ in T so that $e_1, \ldots, e_k, e$ form a circuit, say C. Since $\begin{bmatrix} \epsilon \\ x^C \end{bmatrix}'$ is a $\{0, \pm 1\}$-vector with $M^T \begin{bmatrix} \epsilon \\ x^C \end{bmatrix}' = \mathbf{0}$, it follows that the e-th row of M is a linear combination of the rows $e_1, \ldots, e_k$. Since each of the rows $e_1, \ldots, e_k$ has row sum 0, also row e has row sum 0, i.e., it has exactly one 1 and one -1.

So all rows of M (except for the top row) have exactly one 1 and one -1. This gives us an orientation of G: orient any edge e from v to w if M has a $+1$ in position (e,v) and a -1 in position (e,w).

We show that this is an orientation as required. Let C be a circuit in G. Let $\epsilon = 0$ if $|C|$ is even, and $\epsilon = 1$ if $|C|$ is odd. Now, since

$$M^T \begin{bmatrix} \epsilon \\ x^C \end{bmatrix}' = \mathbf{0} \tag{63}$$

we know that the coordinates e where $\begin{bmatrix} \epsilon \\ x^C \end{bmatrix}'$ is $+1$ and -1, respectively, corresponds to edges e in C oriented one way and the other way, respectively.

Since

$$\left[\begin{bmatrix} \epsilon \\ x^C \end{bmatrix}'\right]^T \begin{bmatrix} 1 \\ x^E \end{bmatrix} = 0 \tag{64}$$

(cf. (61)), it follows that in C the orientation satisfies the condition described in the theorem. $\square$

Gerards showed that Theorem 17 implies the following result of GERARDS and SCHRIJVER [9].

THEOREM 18. *Let $G=(V,E)$ be a graph, without isolated vertices not containing a subgraph isomorphic to the odd-K_4 in (60). Then the coclique polytope of G is equal to the set of all vectors x in $\mathbb{R}^V$ satisfying*

 (i) $x_v \geq 0$ $(v \in V)$,

 (ii) $\sum_{v \in e} x_v \leq 1$ $(e \in E)$,   (65)

 (iii) $\sum_{v \in V(C)} x_v \leq \lfloor \tfrac{1}{2}|V(C)| \rfloor$ (C circuit with $|V(C)|$ odd).

PROOF (sketch). Let P be the set of vectors satisfying (65). Let $G=(V,E)$ be a counterexample to the theorem with $|V|$ as small as possible. First one shows that a minimal counterexample to the theorem should be 3-connected (i.e., there are no two vertices whose removal makes the graph disconnected) — otherwise one could make a smaller counterexample. It is not difficult to check that if a graph is 3-connected and does not contain an odd-K_4, then it neither contains an odd-K_3^2. It follows by Theorem 17 that G can be oriented so that:

 in any circuit, the number of edges oriented one way differs   (66)
 by at most one from the number of edges oriented the other
 way.

Let A denote the set of oriented edges, and let $A^{-1} := \{(v,w)|(w,v) \in A\}$. We now first show the following claim.

CLAIM. *A vector x belongs to P if and only if there exist vectors $y,z \in \mathbb{R}^V$ such that*

 (i) $0 \leq x_v \leq y_v + z_v$ $(v \in V)$,

 (ii) $y_v + z_w \leq 1$ $((v,w) \in A)$,   (67)

 (iii) $y_v + z_w \leq 0$ $((v,w) \in A^{-1})$.

PROOF OF THE CLAIM. 'if': If there exist y,z satisfying (67), condition (i) in (65) is trivial. Condition (ii) holds as for any $\{v,w\} \in E$

$$x_v + x_w \leq (y_v + z_v) + (y_w + z_w) = (y_v + z_w) + (y_w + z_v) \leq 1, \quad (68)$$

since either (v,w) or (w,v) belongs to A.

To check condition (iii), let C be an odd circuit in G. Let $v_0, v_1, \ldots, v_k = v_0$ be a cyclic order of the vertices in C so that $|A \cap \{(v_{i-1}, v_i) | i = 1, \ldots, k\}| = \lfloor \tfrac{1}{2}k \rfloor = \lfloor \tfrac{1}{2}|V(C)| \rfloor$. Then

$$\sum_{v \in V(C)} x_v = \sum_{i=1}^{k} x_{v_i} \leq \sum_{i=1}^{k} (y_{v_i} + z_{v_i}) = \sum_{i=1}^{k} (y_{v_{i-1}} + z_{v_i}) \tag{69}$$

$$\leq |A \cap \{(v_{i-1}, v_i) | i = 1, \ldots, k\}| = \lfloor \tfrac{1}{2} |V(C)| \rfloor.$$

'only if': Define a 'length' function $l: A \cup A^{-1} \to \mathbb{R}$ by:

$$\begin{aligned} l(v,w) &:= 1 - x_v \quad \text{if } (v,w) \in A, \\ l(v,w) &:= -x_v \quad \text{if } (v,w) \in A^{-1}. \end{aligned} \tag{70}$$

Note that each directed cycle C in the directed graph $(V, A \cup A^{-1})$ has nonnegative length $\Sigma_{a \in C} l(a)$, since

$$\sum_{a \in C} l(a) = \sum_{a \in C \cap A} l(a) + \sum_{a \in C \cap A^{-1}} l(a) = - \sum_{v \in V(C)} x_v + |C \cap A| \geq 0, \tag{71}$$

since $|C \cap A| \geq \lfloor \tfrac{1}{2}|V(C)| \rfloor$ by (66) and $\Sigma_{v \in V(C)} x_v \leq \lfloor \tfrac{1}{2}|V(C)| \rfloor$ as $x \in P$ (where $V(C) :=$ set of vertices in C).

Since each directed cycle in $(V, A \cup A^{-1})$ has nonnegative length, there exists a vector $z \in \mathbb{R}^V$ so that $z_w - z_v \leq l(v,w)$ for each $(v,w) \in A \cup A^{-1}$ (we could take $z_v :=$ the minimum length of any directed path in $(V, A \cup A^{-1})$ ending in v - then trivially $z_w \leq z_v + l(v,w)$ for each $(v,w) \in A \cup A^{-1}$). Hence

$$\begin{aligned} z_w - z_v &\leq 1 - x_v \quad \text{if } (v,w) \in A, \\ z_w - z_v &\leq -x_v \quad \text{if } (v,w) \in A^{-1}. \end{aligned} \tag{72}$$

Defining $y_v := x_v - z_v$ we obtain x, y, z satisfying (67). *End of proof of the Claim.*

Now let Q be the set of all vectors $\begin{bmatrix} x \\ y \\ z \end{bmatrix} \in \mathbb{R}^V \times \mathbb{R}^V \times \mathbb{R}^V$ satisfying (67). Then Q is a polyhedron, and it is equal to the convex hull of the integral vectors in Q, i.e., $Q = Q_I$. This follows from the total unimodularity of the constraint matrix in (67), which is of type

$$\begin{bmatrix} I & I & I \\ 0 & M & N \end{bmatrix}, \tag{73}$$

where I is a $V \times V$-identity matrix, and where M and N are $\{0,1\}$-matrices so that every row of M and every row of N contains exactly one 1. By Theorem 5, matrix (73) is totally unimodular, and hence $Q = Q_I$.

Since by the Claim, P is a projection of Q, all vertices of P are integral. Hence, each vertex of P is the characteristic vector of some coclique of G, implying that P is the coclique polytope of G. □

Note that the inequalities (iii) in (65) are the cutting planes added to (i) and (ii). So the theorem states that, if G contains no odd-K_4 as a subgraph, then the coclique polytope is equal to $\{x \in \mathbb{R}_+^V | x_v + x_w \leq 1 \ (\{v,w\} \in E)\}'$.

With the help of Theorem 1 one can derive from Theorem 18 the polynomial-time solvability of the maximum-weighted coclique problem for graphs without odd-K_4. Indeed, one must show that (i), (ii) and (iii) in (65) can be checked in polynomial time. Conditions (i) and (ii) are easily checked one by one. Condition (iii) however consists of exponentially many inequalities. To check them in polynomial time, define a 'length' function $l: E \to \mathbb{R}_+$ by $l(e) := 1 - x_v - x_w$ if $e = \{v, w\}$. Then checking (iii) is equivalent to testing if each odd circuit has length at least 1. This last is not difficult to do in polynomial time, by adaptation of a shortest path algorithm.

GERARDS [8] also derived the following min-max relation.

THEOREM 19. *Let $G = (V, E)$ be a graph without isolated vertices, not containing an odd-K_4 as a subgraph. Then*

$$\max\{|C| \,|\, C \text{ coclique}\} = \min\{|F| + \sum_{i=1}^{t} \lfloor \tfrac{1}{2}|V(C_i)|\rfloor \,|\, F \subseteq E; C_1, \ldots, C_t \quad (74)$$

odd circuits so that $V = \bigcup_{e \in F} e \cup \bigcup_{i=1}^{t} V(C_i)\}$.

So the minimum ranges over all sets of edges and odd circuits which together cover all vertices of G. Note the similarity to Theorem 12.

The theorem means that if we write $\max\{|C| \,|\, C \text{ coclique}\}$ as the problem of maximizing $\mathbf{1}^T x$ over vectors x satisfying (65), we obtain a linear program with integral optimum primal *and* dual solutions.

FINAL REMARK. In this paper we saw the polynomial-time solvability of two combinatorial optimization problems:

(i) finding a maximum-weighted matching in a graph; (75)
(ii) finding a maximum-weighted coclique in a graph without odd-K_4.

In fact, a maximum-weighted matching in a graph H can be considered as a maximum-weighted coclique in the line-graph $L(G)$ of G. MINTY [17] and SBIHI [19] showed that more generally, the maximum-weighted coclique problem for *claw-free* graphs is polynomially solvable, i.e., for graphs not containing

as an *induced* subgraph. By Theorem 1 it implies that the separation problem for coclique polytopes of claw-free graphs is polynomially solvable. However, no explicit description by inequalities for these polytopes has been found. It has been shown by CHVÁTAL [3] that there exists no fixed t so that for claw-free graphs the coclique polytope is equal to $\{x\in\mathbb{R}_+^V | x_v+x_w \leq 1 (\{v,w\}\in E)\}^{(t)}$, in the notation of Section 7.

REFERENCES
1. C. BERGE (1958). Sur le couplage maximum d'un graphe. *Comptes Rendus Hebdomadaires des Séances de l'Académie des Sciences 247*, 258-259.
2. V. CHVÁTAL (1973). Edmonds polytopes and a hierarchy of combinatorial problems. *Discrete Mathematics 4*, 305-337.
3. V. CHVÁTAL (1984). *Cutting-Plane Proofs and the Stability Number of a Graph,* Report No. 84326, Institut für Operations Research, Universität Bonn.
4. S.A. COOK (1971). The complexity of theorem-proving procedures. *Proceedings of the Third Annual ACM Symposium on Theory of Computing,* The Association of Computing Machinery, New York, 151-158.
5. J. EDMONDS (1965). Maximum matching and a polyhedron with 0,1-vertices. *Journal of Research of the National Bureau of Standards (B) 69*, 125-130.
6. A. FRANK, É. TARDOS (1985). An application of simultaneous approximation in combinatorial optimization. *26th Annual Symposium on Foundations of Computer Science,* IEEE, New York, 459-463.
7. A.M.H. GERARDS. *Homomorphisms of Graphs into Odd Cycles,* preprint.
8. A.M.H. GERARDS, to appear.
9. A.M.H. GERARDS, A. SCHRIJVER. Matrices with the Edmonds-Johnson property, to appear in *Combinatorica.*
10. R.E. GOMORY (1958). Outline of an algorithm for integer solutions to linear programs. *Bulletin of the American Mathematical Society 64*, 275-278.
11. M. GRÖTSCHEL, L. LOVÁSZ, A. SCHRIJVER (1981). The ellipsoid method and its consequences in combinatorial optimization. *Combinatorica 1*, 169-197.
12. M. GRÖTSCHEL, L. LOVÁSZ, A. SCHRIJVER (1986). *The Ellipsoid Method and Combinatorial Optimization,* Springer-Verlag, to appear.
13. A.J. HOFFMAN, J.B. KRUSKAL (1956). Integral boundary points of convex polyhedra. H.W. KUHN, A.W. TUCKER (eds.). *Linear Inequalities and Related Systems,* Princeton University Press, Princeton, N.J., 223-246.
14. R.M. KARP (1972). Reducibility among combinatorial problems. R.E. MILLER, J.W. THATCHER (eds.). *Complexity of Computer Computations,* Plenum Press, New York, 85-103.
15. L.G. KHACHIYAN (1979). A polynomial algorithm in linear programming (in Russian). *Doklady Akademii Nauk SSSR 244*, 1093-1096 (English translation: *Soviet Mathematics Doklady 20*, 191-194).
16. A.K. LENSTRA, H.W. LENSTRA, JR., L. LOVÁSZ (1982). Factoring polynomials with rational coefficients. *Mathematische Annalen 261*, 515-534.

17. G.J. MINTY (1980). On maximal independent sets of vertices in claw-free graphs. *Journal of Combinatorial Theory (B) 28,* 284-304.
18. M.W. PADBERG, M.R. RAO (1982). Odd minimum cut-sets and b-matchings. *Mathematics of Operations Research 7,* 67-80.
19. N. SBIHI (1980). Algorithme de recherche d'un stable de cardinalité maximum dans un graphe sans étoile. *Discrete Mathematics 29,* 53-76.
20. A. SCHRIJVER (1980). On cutting planes. *Annals of Discrete Mathematics 9,* 291-296.
21. A. SCHRIJVER (1983). Short proofs on the matching polyhedron. *Journal of Combinatorial Theory (B) 34,* 104-108.
22. A. SCHRIJVER (1986). *Theory of Linear and Integer Programming,* Wiley, Chichester.
23. P.D. SEYMOUR (1977). The matroids with the max-flow min-cut property. *Journal of Combinatorial Theory (B) 23,* 189-222.
24. P.D. SEYMOUR (1980). Decomposition of regular matroids. *Journal of Combinatorial Theory (B) 28,* 305-359.
25. P.D. SEYMOUR (1981). Matroids and multicommodity flows. *European Journal of Combinatorics 2,* 257-290.
26. W.T. TUTTE (1947). The factorization of linear graphs. *Journal of the London Mathematical Society 22,* 107-111.
27. W.T. TUTTE (1958). A homotopy theorem for matroids I, II. *Transactions of the American Mathematical Society 88,* 144-174.

Archirithmics or Algotecture?

Paul M.B. Vitányi

Massachusetts Institute of Technology
Laboratory for Computer Science
Cambridge, Massachusetts
and
Centre for Mathematics and Computer Science
P.O. Box 4079, 1009 AB Amsterdam, The Netherlands

The interaction between algorithms, and the architecture of the machines on which they are executed, shapes computer science. Currently the emphasis in computer science is rapidly shifting from serial computing to parallel and distributed computing. In sequential computing we can ignore many details of the underlying physical machine to obtain a clean computation model and nice and 'mathematical' algorithms. In contrast, in parallel or distributed computing we often have to reckon with many physical details of the actual computing complex—thus obtaining an opaque computation model at best. This is illustrated by the inherent problems of communication and 'wires' in multiprocessor systems. Another aspect of distributed systems is the emergence of a category of algorithmic problems which make no sense in the context of sequential computation: problems of distributed control.

> 'You see, it's like a portmanteau—there are two meanings packed up in one word'
> Lewis Carroll

1. ARCHITECTURE + ALGORITHMS ≈ COMPUTING

The earliest electronic computing engines arose as a byproduct of the Manhattan Project in World War II. Broadly speaking, their purpose was to compute numerical solutions to second order partial differential equations arising in connection with the design of the atomic bomb. The machines consisted of primitive logical and memory components like electromagnetic relays and mercury delay lines, which where wired up so as to have the complex perform the desired computation. The *architecture* reflected the type of *algorithm* to be performed, namely for solving the equations mentioned by numerical grid methods. Such algorithms suggest parallel or pipelined execution, and that is exactly the type of architecture of those first computers [1]. Only at the

---

1. This work was supported in part by the Office of Naval Research under Contract N00014-85-K-0168, by the Office of Army Research under Contract DAAG29-84-K-0058, by the National Science Foundation under Grant DCR-83-02391, and by the Defense Advanced Research Projects Agency (DARPA) under Contract N00014-83-K-0125.

present time, in the middle eighties, have we come full circle and see such special purpose architectures again in the pipelined and systolic algorithms frozen in the silicon hardware of chips. Once more, the shift is away from sequential thinking in the form of line-by-line programs of imperative or other nature, and to representing algorithms in structures of space and time.

After the Manhattan Project had been fulfilled, project leader J.R. Oppenheimer stated: 'It flashed through my mind that I had become the Prince of Darkness, the Destroyer of Universes' [Hindu epic *Bhagavad Gita*]. This earned him the dour rejoinder: '[he] professes guilt, to claim credit for the sin' from mathematician J. von Neumann. He [JvN] and other computer designers quickly progressed to the idea of automating all types of computational tasks. Rather than stooping to the chore of rewiring a new complex for every new task which came along, the idea arose of letting the computer take over that job as well. Thus, the idea of a *general purpose* computer entered the scene. It so happened that mathematicians like H.H. Goldstine, J. von Neumann and A.W. Burks were well aware of A.M. Turing's brilliant 1936 paper [2] in which he described an architecture for just such a hypothetical machine.

> "Computing is normally done by writing certain symbols on paper. We may suppose this paper to be divided into squares like a child's arithmetic book. In elementary arithmetic the two-dimensional character of the paper is sometimes used. But such use is always avoidable, and I think that it will be agreed that the two-dimensional character of paper is no essential of computation. I assume then that the computation is carried out on one-dimensional paper, i.e., on a tape divided into squares. I also suppose that the number of symbols which may be printed is finite. The behaviour of the [human] computer at any moment is determined by the symbols he is observing, and his 'state of mind' at that moment. We may suppose that there is a bound B to the number of symbols or squares which the computer can observe at one moment. If he wishes to observe more, he must use successive observations. We will also suppose that the number of states of mind which need be taken into account is finite.
>
> We suppose [above] that the computation is carried out on a tape; but we avoid introducing the 'state of mind' by considering a more physical and definitive counterpart of it. It is always possible for the computer to break off from his work, to go away and forget all about it, and later to come back and go on with it. If he does this he must leave a note of instructions (written in some standard form) explaining how the work is to be continued. This note is the counterpart of 'the state of mind'. We will suppose that the computer works in such a desultory manner that he never does more than one step and writes the next note. Thus the state of progress of the computation at any stage is completely determined by the note of instructions and the symbols on the tape. That is, the state

of the system may be described by a single expression (sequence of symbols), consisting of the symbols on the tape followed by Δ (which we suppose not to appear elsewhere) and then by the note of instructions. This expression may be called the 'state formula'. We know that the state formula at any given stage is determined by the state formula before the last step was made, and we assume that the relation of these two formulae is expressible in the functional calculus. In other words, we assume that there is an axiom A which expresses the rules governing the behaviour of the computer, in terms of the relation of the state formula at any stage to the state formula at the preceding stage. If this is so, we can construct a machine to write down the successive state formulae, and hence to compute the required number."

Grasping the implied architectural concept, and improving it according to the leeway provided by physical law, BURKS, GOLDSTINE and VON NEUMANN in 1946 wrote a memorandum [3] which shaped the architecture of electronic computers for the next forty years. This memorandum was preceded by the famous 'First Draft' [4], were we can clearly distinguish the serial mode of operation of the modern computer, i.e., one instruction at a time is inspected and then executed. This is in sharp distinction to the parallel operation of the earlier ENIAC computer in which many things were simultaneously being performed. To abandon all parallelism was not thought of as detrimental to performance, since the potential speed of the electronic techniques was judged to be fast enough. Complainants about the 'Von Neumann' bottleneck (explained below), inherent in the stored program sequential computer as we know it, should realize that the conceptual advantage of this scheme is what made possible the giant strides of progress: if cars had become so much cheaper as computing power has, a car would cost less than 1 dollar. Turing's analysis of the process of computation as the sequential execution of a sequence of operations is so natural, that it seems as if Euclid in designing one of the earliest known algorithms (for computing the greatest common divisor) must have had such an architecture in mind. Yet at the moment it seems that the heyday of sequential computing is rapidly fading; and various forms of distributed computing are on the rise. Simultaneously, the stress on the design of sequential algorithms is rapidly shifting to non-sequential ones. We want to focus on the heart of Computing as this interplay between Architecture and Algorithms, the one aspect shaping the other in a continuous interaction: *Archirithmics* or *Algotecture*. We shall contrast the *paradise* of sequential computation with the *jungle* of non-sequential computing. In sequential computing we can ignore many details of the underlying physical machine to obtain a clean computation model and nice and 'mathematical' algorithms. In contrast, in parallel or distributed computing we often have to reckon with many physical details of the actual computing complex - thus obtaining an opaque computation model at best. This is illustrated by the inherent problems of communication and 'wires' in multiprocessor systems. Another aspect of distributed systems is the emergence of a category of algorithmic problems which make no sense in the context of sequential computation: problems of distributed control. But it is fitting here to look into the development of Algotecture at the Mathematical Centre first.

2. CWI AND THE RISE OF ARCHIRITHMICS[1]

The Mathematical Centre was founded in 1946, just after World War II, in Amsterdam. The young mechanical engineer A. van Wijngaarden, having turned to numerical mathematics earlier, was one of the first employees. His background combined sense of architecture ('architect' is literally, 'master of building') with sense of algorithms from numerical mathematics, and he therefore was an appropriate figure to become the father of computing science in the Netherlands. In the first two months of 1946, being in England, he marveled at the miracles being wrought there. His was the task to build a computing department at the new Mathematical Centre, which was supposed to include the building of computing engines. Traveling in England and the United States in 1947, he visited von Neumann's group in Princeton, and, subsequent to engaging two physics students (B.J. Loopstra and C.S. Scholten): 'after buying a screwdriver we tried to build an integrator'. Being out of funds and other support, the group started with - or so they thought - a moderate design, that of a relay computer. This ARRA (Automatische Relais Rekenmachine Amsterdam) was heavier than anticipated, according to the 1949 yearbook of the Mathematical Centre: 'The brick and mortar [building] cause great anxiety. It takes art and improvisation to prevent the wooden floor collapsing, while no precision toolwork can be performed because of lack of adequate foundations.' In 1950 the machine performed its first calculations, and in 1952 it was officially inaugurated. While the machine broke down, producing an incoherent sequence of digits, demonstrator Van Wijngaarden explained with great acumen that now the device performed a most difficult task of producing a genuine random sequence. (So I have been told. Others claim that the wily demonstrator, expecting a breakdown, chose the generation of a pseudo-random sequence as both a *difficult* task, and a task that could *hardly fail*.) In 1952 student E.W. Dijkstra joined the crew, and in 1953 a new, electronic, version of the ARRA came into service. This machine was a purely sequential computer manufactured with plug-in components. While a copy of the ARRA with larger memory was built for Fokker aircraft industry, at the Mathematical Centre a new and much faster machine, the ARMAC (Automatische Rekenmachine Mathematisch Centrum), was developed and taken into service in 1956. The ARMAC had - in contrast to its predecessors - parallel data paths and already used - some - ferrite core memory and transistors. In the same year 1956 that the ARMAC replaced the ARRA, the Mathematical Centre started developing a fast machine built with magnetic core memory, diodes and transistors, called the X1. Two things were genuine new features in the

---

1. We neglect the remainder of the Netherlands here. The interested reader should consult [5].

X1: the use of solid-state components throughout, compactly packaged on boards in closed housings with edge connectors, so that defective gates could be replaced as defective pc boards are replaced today (anticipated by the plug-in components of ARRA II and ARMAC), and the facility of a dynamic interrupt mechanism. This latter feature had been installed, at the insistence of A.W. Dek and after having been polished by E.W. Dijkstra, and was named 'ingreep' from which the later English word 'interrupt' was derived. The development of the X1 led to an agreement with the startup of Electrologica NV to transplant the entire computer building and development activities at the Mathematical Centre to that company. In 1958 the Mathematical Centre finally stopped computer building altogether. While the hardware activities thus disappeared, the software activities were on the rise. Van Wijngaarden was one of the authors of the Algol 60 Report, defining the prototype advanced programming language. Within half a year from publication, E.W. Dijkstra and J.A. Zonneveld completed the first - and also exemplary - Algol 60 compiler. These historical facts and more can be found in [5], a publication produced by means of an early text formatter, written by H. Brandt Corstius in Algol 60 at the Mathematical Centre, run on the Electrologica X1 there. The formatter included an advanced hyphenation algorithm for Dutch words and text filling with fractional spacing. About eight years later (or more), the ultimate programming language Algol 68 was defined, using concepts and methods which had been primarily developed at the Mathematical Centre, mainly due to Van Wijngaarden.

On February 11, 1966, at the 20th Mathematical Centre anniversary, Director Van Wijngaarden received a box containing binoculars saying 'magnifies 8X', which symbolized a necessary new room to house the X8 computer. Now the MC has changed its name to CWI (Centre for Mathematics and Computer Science), the housing of the computers is not the greatest problem anymore - they get smaller fast enough to compensate for their increase in number - but rather how to house the growing number of humans.

3. GENEALOGY

In a recent issue of SIGACT News, D.S. Johnson of AT&T Bell Laboratories in Murray Hill (N.J.) started a compilation of the *genealogy* of theoretical computer science. The rules are that the genealogy is a network consisting of nodes representing computer scientists. There is an arc from node A to node B if A got his Doctorate under B. (B has to be the official Thesis Adviser or 'Promotor'.) Contrary to ordinary genealogy, although a father may have many sons, he usually has but *one* direct ancestor. In fact, this genealogy is a genuine tree (rather, a forest), like the paternal (or maternal) ancestor tree. As a contribution to the sociology of the Dutch part of the forest, I like to drop a seed here - without claiming that the arbitrary selection below is representative.[1]

---

1. That is left to a qualified person who also has better access to the archives than I have now. P. van Emde Boas is compiling a large genealogy for publication. Those who want to be included, or have information about entries which ought to be included, are encouraged to send a note to Dr. P. van Emde Boas at the Informatica Department of the University of Amsterdam.

Advisee	Advisor	University	Year
L.E.J. Brouwer	J. Korteweg	Univ. of Amsterdam	1907
A. Heyting	L.E.J. Brouwer	Univ. of Amsterdam	1925
A. van Wijngaarden	C. Biezeno	Tech. Univ. Delft	1945
E.W. Dijkstra	A. van Wijngaarden	Univ. of Amsterdam	1959
D. van Dalen	A. Heyting	Univ. of Amsterdam	1963
J.W. de Bakker	A. van Wijngaarden	Univ. of Amsterdam	1969
J. van Leeuwen	D. van Dalen	Univ. of Utrecht	1972
P. van Emde Boas	A. van Wijngaarden	Univ. of Amsterdam	1974
W.P. de Roever	J.W. de Bakker	VU Amsterdam	1975
J.K. Lenstra	G. de Leve	Univ. of Amsterdam	1976
M. Rem	E.W. Dijkstra	Tech. Univ. Eindhoven	1976
H.W. Lenstra, Jr.	F. Oort	Univ. of Amsterdam	1977
P.M.B. Vitányi	J.W. de Bakker	VU Amsterdam	1978
M. Overmars	J. van Leeuwen	Univ. of Utrecht	1983
A.K. Lenstra	P. van Emde Boas	Univ. of Amsterdam	1984
S.J. Mullender	A.S. Tanenbaum	VU Amsterdam	1985

What strikes the eye here is that the trees have roots. A.S. Tanenbaum is a root because his ancestor is not included in the table. Many Dutch computer scientists belong to the Biezeno tree. C. Biezeno has no ancestor because he is not a genuine Doctor but a *Doctor Honoris Causa*: the family tree ends there. The Korteweg tree is an honourable tree. J. Korteweg has no ancestor because, notwithstanding the genuine degree, we have been unable to identify a definite *promotor*. But let us proceed with the main subject of this paper.

4. Sequential computation

In sequential computation one can ignore many physical details of the underlying computer system in analysing the computational complexity of some program. Each operation essentially consists of a sequence of 'fetch from memory', 'execute operation on one or more operands in the Central Processing Unit' and 'store in memory'. The CPU operations can be thought of - when viewed from sufficient distance - as essentially finite automata transitions which transform input obtained by a bounded number of 'fetch from memory' operations (say 2) into output in the form of 'store in memory' operations (say 1). In the usual setup, a memory register has a fixed length (say 48 bits) and both the memory accesses and CPU operations take a fixed time (say, at most X). Therefore, a sequence of n operations takes in between nX and $4nX$ time. Forgetting about the X and the small constants like 4, it is usual to say that n operations take n 'time'. Note, here 'time' means number of steps. Similarly, it is assumed that all objects manipulated fit in a single memory location. Moreover, that each object is 'randomly accessible', that is, each object can be accessed as fast as any other. This is referred to as the 'unit cost measure'.

This scheme is sometimes refined to take into account that some items being manipulated do not fit in a 48 bit register - for instance the 123rd Mersenne prime. It is then customary to charge the cost of manipulating the item as being linear in its length, both in terms of storage and in terms of time for execution of an operation. This is referred to as the 'logarithmic cost measure'. It is clear that this time cost measure is only a lower bound since the actual operations performed on the items when chopped up, often require super-linear time in the length of the items. For instance, while logarithmic cost may be reasonable for addition, it is not reasonable for multiplication.

A further refinement may be made for objects not held in 'random access' memory, but on disk or mass storage devices such as tapes. There an operation on an object may involve swapping pieces of the object back and forth from disk to random access memory, thus incurring a time overhead which may be orders of magnitudes greater than the time spent on manipulating in the CPU and random access memory. Think about the sorting or merging of huge data files. The logarithmic cost measure tries to take such an overhead into account by charging as the cost of a memory access also the length of the memory address. As in the case of the registers, this can be only a very crude lower bound on the actual cost. We thus distinguish a memory hierarchy, where the access times of objects stored at different levels differ orders of magnitudes.

While the physical aspects of computing devices can thus be fairly well accounted for, the basic unit of time a transaction takes does not vary too wildly within each level we have distinguished. It is therefore more or less justified to forget about the details and talk only about the number of operations at each level of the memory hierarchy. A reasonably valid model of a processing environment, where we ignore many of the real life aspects of the device(s), is called a *computational model*. In the design and analysis of algorithms for sequential computers, the computational model can be clean and abstract. That is, we can profitably proceed without too much knowledge of the underlying physical details of the computing complex. As we will see, in the realm of non-sequential computation reality can not be ignored to such an extent.

Since in current computers the time of a basic operation in the CPU is generally far lower than that of memory accesses, most computations are memory bound, i.e., the time spent in accessing various levels in the memory hierarchy completely dominates the computation time. This is popularly called the 'Von Neumann' bottleneck. Are the prospects any brighter in the coming era of non-sequential computation?

5. SPACE AND COMMUNICATION

In many areas of the theory of parallel computation we meet graph-structured computational models. These models suggest the design of parallel algorithms in which the cost of communication is largely ignored. Yet it is well known that the cost of computation - in both time and space - vanishes with respect to the cost of communication in parallel or distributed computing. As multiprocessor systems with really large numbers of processors start to be constructed, this effect becomes more and more apparent. Thinking Machines Corporation of Cambridge, Mass., has just marketed the 'Connection Machine', a massively multiprocessor parallel computer. The prototype contains microscopically fine grained processor/memory cells, 65,536 of them, each with 4,096 bits of memory and a simple arithmetical unit. The communication network connecting the processors is packet switched and based on the binary n-cube[1] ($n=16$). The processors execute a single stream of instructions generated by a microcontroller under direction of a conventional host. The machine is packaged in a cube with sides of 1.3 meters. Some specifications are as follows. Total memory $2,5 \times 10^8$ bits; memory bandwidth 2×10^{11} bits/second; processor bandwidth 3×10^{11} bits/second; communication bandwidth 3×10^7 worst case up to 5×10^{10} bits/second; and input/output bandwidth 5×10^8 bits/second. The basic operation of a processing element reads two bits from the associated memory plus one flag and combines them according to a specified logical operation to produce one bit for the memory and one flag bit. Larger operands are treated by many processing elements working in concert. In conventional terms the machine has a peak of 1000 million instructions per second (of 32-bits additions). It is air cooled and dissipates 12,000 W running on a 4 MHz clock. This design does not have a processor/memory (von Neumann) bottleneck. However, it *does* have a wiring (Non-von Neumann) bottleneck as we see below. To alleviate the wiring up of such a machine, the makers settled for a 2^{16}-processor prototype. By designing a chip to contain 16 processor cells and one router unit of the communication network, they got rid of an exponent 4 at the bottom. The communications network in the Connection Machine is formed by the 4,096 routers connected by 24,576 bidirectional wires in the pattern of the binary 12-cube. 'New Computer Architectures and their Relationship to Physics or, Why Computer Science is No Good', the last chapter of [6], expresses the dissatisfaction of the designers with traditional computer science, 'which abstracts the wire away into a costless and volumeless idealized connection. [The] old models do not impose a locality of connection, even though the real world does. ... In classical computation the wire is not even considered. In current engineering it may be the most important thing'. Below we analyse some recent delinquent theoretical models for parallel computation which suffer from *serialitis*. That is to say, they do not deal with the intrinsic communication problems associated with geographically distributed computing, and therefore may suggest that algorithms are good which in reality are bad.

(1) For example, 'parallel random access machines (PRAM's)' can at each point in their computation spawn a couple of offspring PRAM's to perform some subcomputations. Broadly speaking, we can therefore imagine the computation as a binary tree of processors. The 'time' the computation takes is then linearly related to the depth of the tree.

---

1. A network with 2^n nodes, each node identified by an n-bit name. There is an edge between nodes that differ in a single bit.

(2) In [7] this idea is translated into terms of 'very large integrated circuits'. In Chapter 8 the authors show a bold picture of a complete binary tree, and explain that such a tree with processors in each node, is capable of solving NP-complete problems like the 'traveling salesman problem' in linear time. This, on the grounds that the processor at the root can send a copy of the problem instance to each of the leaves, and each of the leaves can try one candidate solution. A simple scheme can guarantee that each leaf tries a different solution, each solution is tried by some leaf, and all answers are percolated upwards to the root. If positive answers win over negative ones in the fan-in, the answer the root receives is a solution if there is one and 'no solution' if there is none.

(3) One of the currently flourishing parts of the theory of parallel computation is 'NC-computation'. A problem is in 'Nick's Class' if it can be solved[1] in polylogarithmic 'time' using a polynomial number of processors. Here, 'time' means the length of the longest chain of causally related steps.

All of the above models may say something about the parallelizability of algorithms for certain problems. This often takes the form of distributing copies of the entire problem instance, or pieces of the problem instance, among an exponential number of processors in a linear number of steps. Or, as in NC, among a polynomial number of processors in a polylogarithmic number of steps. The way a problem instance can be divided and partial answers put together may give genuine insight into its parallelizability. However, it can *not* give a reduction from an asymptotic exponential time best algorithm in the sequential case to an asymptotic polynomial time algorithm in *any* parallel case. At least, if by 'time' we mean time. This can be seen easily as follows. If the parallel algorithm uses 2^n processing elements, regardless of whether the computational model assumes bounded fan-in and fan-out or not, it can not run in time polynomial in n, because *physical space* has us in its tyranny. For, if we use 2^n processing elements of, say, unit size each, then the tightest they can be packed is in a 3-dimensional sphere of volume 2^n. No unit in the sphere can be closer to all other units than a distance of radius R,

$$R = \left[\frac{3 \cdot 2^n}{4\pi}\right]^{1/3}$$

Modulo a major advance in physics, it is impossible to transport signals over $2^{\alpha n}$ ($\alpha > 0$) distance in polynomial $p(n)$ time. In fact, the assumption of the bounded speed of light suggests that the lower time bound on *any* computation using 2^n processing elements is $\Omega(2^{n/3})$ outright. Or, for the case of NC-computations which use n^α processors, $\alpha > 0$, the lower bound on the computation time is $\Omega(n^{\alpha/3})$.[2]

---

1. Named after Nicholas Pippenger.
2. It is sometimes argued that this effect is significant for large values of n only, and therefore can safely be ignored. This is a curious defense in an area where all results are of asymptotic nature, i.e., hold only for large values of n.

The situation is worse than it appears on the face of it. Let us analyse the amount of wire involved. To prevent arguments that the results hold only asymptotically, or that processors are huge and wires thin, we calculate precisely without hidden constants and assume that wires have length but no volume and can pass through everything. Consider an architecture such as the binary n-cube. Recall, that this is the network with $N=2^n$ nodes, each of which is identified by an n-bit name. There is a communication edge between two nodes if their identifiers differ in a single bit. Call this graph $C=(V,E)$. Let C be embedded in 3-dimensional Euclidean space, and let each node have unit volume. Let x be any node of C. There are at most $2^n/8$ nodes within Euclidean distance $R/2$ of x, where R is as above. Then, there are $\geq 7\cdot 2^n/8$ nodes at Euclidean distance $\geq R/2$ from x. Construct a spanning tree T_x of C of depth $\leq n$ with node x as the root. The average Euclidean length of a path from the root in T_x is $\geq 7R/16$, and therefore the average Euclidean length of an embedded edge in a path from the root in T_x is $\geq 7R/16n$. This does not give a lower bound on the average Euclidean length of an edge in T_x. However, using the symmetry of the binary n-cube we can establish that the average Euclidean length of the edges in the 3-space embedding of C is $\geq 7R/16n$. We can prove this as follows.

PROOF. Denote a node a in C by an n-bit string $a_1 a_2 \cdots a_n$, and an edge (a,b) between nodes a and b differing in the ith bit by:

$$a = a_1 \cdots a_{i-1} \hat{a}_i a_{i+1} \cdots a_n$$

This means that an edge has two representations. Now we can express a set I of isomorphic mappings of C to itself by (1) a cyclic permutation of the representation of nodes and edges, followed by (2) complementation of the bits of the representations in a given pattern. I.e., the isomorphism $(j, c_1 c_2 \cdots c_n) \in I$ maps the above edge a to

$$b = b_{j+1} \cdots b_{i-1} \hat{b}_i b_{i+1} \cdots b_n b_1 \cdots b_j$$

with $b_i = a_i$ if $c_i = 0$ and $b_i = \overline{a_i}$ ($=$ complement a_i) if $c_i = 1$. Consider the ensemble S of spanning trees of C, each tree isomorphic with T_x above, consisting of the $n2^n$ trees $i(T_x)$ to which T_x is mapped by the $n2^n$ distinct isomorphisms i in I. For each edge e in T_x and each edge e' of C there are two distinct isomorphisms i_1 and i_2 in I such that $i_1(e) = i_2(e) = e'$. The average Euclidean length of a path from the root in each tree $i(T_x) \in S$ ($i \in I$) is $\geq 7R/16$, so the average Euclidean length of a path from the root taken over all trees $i(T_x) \in S$ ($i \in I$) is $\geq 7R/16$ as well. Let the Euclidean length of an edge e in the 3-space embedding of C be $l(e)$. Then, for each edge e of T_x:

$$\sum_{i \in I} l(i(e)) = 2 \sum_{e \in E} l(e)$$

That is, each edge in the embedded C occurs twice as the same edge of the canonical tree T_x in the form of the corresponding isomorphic edge in some tree in S. Therefore, the average Euclidean length of the edges in trees in S, which correspond to a single particular edge of T_x, equals the average Euclidean length of an edge in E. Let P be a path from the root in T_x consisting of $|P| \leq n$ edges. Then, the average sum of the Euclidean lengths of the edges in a path $i(P)$ from the root in all trees $i(T_x)$ ($i \in I$) equals $|P|$ times the average Euclidean edge length in E:

$$\sum_{e \in P} \sum_{i \in I} l(i(e)) = 2|P| \sum_{e \in E} l(e)$$

Consequently, the average Euclidean edge length in E equals the average Euclidean length of an edge in a path P from the root in a tree in S, and is therefore $\geq 7R/16n$:

$$\frac{1}{n 2^{n-1}} \sum_{e \in E} l(e) = \frac{1}{2^n} \sum_{P \in T_x} \frac{1}{n 2^n |P|} \sum_{e \in P} \sum_{i \in I} l(i(e)) \geq \frac{7R}{16n}$$

Since there are $n 2^n/2$ edges in the binary n-cube, this sums up to an amazing *total* wire length $\sum_{e \in E} l(e)$ needed in the Euclidean 3-dimensional embedding of C of

$$\sum_{e \in E} l(e) \geq \frac{2^n 7R}{32} \geq \left(\frac{3}{4\pi}\right)^{1/3} \cdot 7 \cdot 2^{(4n/3)-5}$$

Many network topologies are afflicted with this problem: n-dimensional cube networks, Fast Fourier Transform (FFT) networks, butterfly networks, shuffle-exchange networks, cube-connected cycles networks, and so on. In fact, the arguments seem to hold for networks with a small diameter which satisfy certain symmetry requirements. An example of a network with small diameter which is not symmetric in this sense is the tree. The fact that 7/8th of all paths from the root in a complete tree would have Euclidean length $\geq R/2$ in a 3-space embedding does not imply that the average Euclidean length of an embedded edge of the tree is larger than a constant. This is borne out by the familiar H-tree layout [7] where the average edge length is less than 3 or 4.

NOTE. Deriving the result about the total necessary wire length for embedding the binary n-cube, we did not make *any* assumptions about the volume of a wire of unit length, or the way they are embedded in space, as is usual [8]. It is consistent with the derived results that wires have *zero* volume, and that *infinitely* many wires can pass through a unit 2-dimensional area. Such assumptions invalidate the arguments used elsewhere. In contrast to other investigations, the goal here is to derive lower bounds on the total wire length, irrespective of the ratio between the volume of a unit length wire and the volume of a processing element. The lower bound on the total wire length above is independent of this ratio, which changes with different technologies or granularity of computing components.

Iterating this reasoning, but now adding the volume of the wires to the volume of the nodes, the greatest lower bound on the volume necessary to embed the binary n-cube converges to a particular solution in between a total volume of $\Omega(2^{4n/3})$ and a total volume of, say, $O(2^{2n})$ if we charge a constant fraction of the unit volume for a unit wire length. The lower bound $\Omega(2^{4n/3})$ ignores the fact that the added volume of the wires pushes the nodes further apart, thus necessitating longer wires again. The $O(2^{2n})$ upper bound holds under the assumption that wires of all lengths have the same volume per unit length (not more than a constant fraction of the unit volume of a node). In [9,10] it is shown that the latter assumption cannot always be made. (For instance, if we want to drive the signals to very high speed on chip.)

More in general we can say the following.

DEFINITION (sketch). *Call a network isotropic if all edges are 'symmetric' in the above sense.*

THEOREM. *Let R_d be the radius of a d-dimensional sphere of volume N. Let G be an N-node isotropic network. Let D be the diameter of G. Let a d-dimensional embedding of G in Euclidean d-dimensional space be such that each node has volume 1. Assume that a node is a sphere and not a 'funny' form like a wire. Allow that wires have no volume and can cross through nodes in arbitrary ways. The average Euclidean length of an embedded edge in such an embedding of G is $\geq (2^d-1)R_d/(2^{d+1}D)$. For the 3-dimensional embedding of a complete graph K_N this results in an average wire length of $\geq 7R_3/16$, with $R_3 = (3N/4\pi)^{1/3}$. For the 3-dimensional embedding of an N-node ring this results in an average wire length of $\geq 7R_3/8N$. (Let $N > 100$, say.)*

These surprising facts are a theoretical prelude to many *wiring problems* currently starting to plague computer designers and chip designers alike. Formerly, a wire had magical properties of transmitting data 'instantly' from one place to another (or better, to many other places). A wire did not take room, did not dissipate heat, and did not cost anything - at least, not enough to worry about. This was the situation when the number of wires was low, somewhere in the hundreds. Current designs use many millions of wires (on chip), or possibly billions of wires (on wafers). In a computation of parallel nature, most of the time seems to be spent on communication - transporting signals over wires. Thus, thinking that the von Neumann bottleneck has been conquered by non-sequential computation, we are unaware that the Non-von Neumann bottleneck is still waiting. The following innominate quote covers this matter admirably:

'Without me they fly they think;
But when they fly I am the wings.'

It is clear that these communication mishaps have influence on the algorithms to be designed for the massive multiprocessors of the future, and, *vice versa,* existing algorithms influence the creation of novel architectures (e.g., the k-ary n-cube Mosaic of Caltech, the FFT Butterfly of Bolt, Beranek and Newman, the shuffle-exchange Ultracomputer of New York University) to run them on.

Another effect which becomes increasingly important is that most of the room in the device executing the computation is taken up by the wires. Under the very conservative estimate that the unit length of a wire has a volume which is a constant fraction of that of a component it connects, we can see above that in 3-dimensional layouts for binary n-cubes, or for the other fast permutation networks, the volume of the 2^n components performing the actual computation operations is, asymptotically, a quickly vanishing fraction of the volume of the wires needed for communication:

$$\frac{\text{volume computing components}}{\text{volume communication wires}} \in o(2^{-n/3})$$

Today it seems that a partial solution to this problem may be found in optical communication, either wireless by means of lasers/infrared light or by using virtually unlimited bandwidth optical fiber or integrated waveguides [11]. But beware, even while Nature is not malicious, she is subtle.

6. Distributed control

It is useful to distinguish between *distributed computation* and *distributed control*. The former is also concerned with the distributed solution of problems for which there also exist sequential algorithms, that is, algorithms which are in essence only distributed in time. Examples are parallel algorithms for matrix multiplication, FFT, graph problems, and so on. Distributed control is concerned with problems which make no sense in terms of sequential computation. The subject matter of such problems is the *organization* of computation distributed in *both* space and time. The distribution in space can be on a geographically small scale, like on a chip, on a larger scale in a multiprocessor system, or in local area networks on campus, or worldwide wide area networks. Each of these categories has its own special problems, and there are problems which are common to all. Rather than exhaustively enumerate the many such issues which are the current object of intensive study and grant proposals all over the globe, I will select two recent examples, which I think give a flavour of the field.

6.1. Distributed match-making

Suppose you want to give a party in your Silicon Valley home, but do not care for the bother. You want a catering service. Now it so happens that you do not know the address or telephone number of such a service. Anyway, even if you did, this would not do you much good. In Silicon Valley such small outfits come and go so fast that it is unlikely that this service, which you used two years ago, still exists at the old address. You can phone them, but the number gets you somebody who has never heard of your old catering service. There are several courses of action you can take.

- One way to solve your problem is to send mail to everybody in town asking whether they supply catering service. In computer networks this is called *broadcasting*.
- Another way is to wait until you get an advertisement leaflet of a catering service in your mailbox. Below we call this *sweeping*.

Most likely, you do one of the following:
- You look in the Yellow Pages under the appropriate heading. If everybody exclusively uses YP for all services then we may view the YP outfit as a centralized name server. Services reveal their whereabouts by advertising there and clients look them up there. If the YP company crashes then clients and services cannot be matched anymore, and society grinds to a halt.
- You buy a suitable newspaper and look up 'catering' in the advertisement section. Now the name server is distributed. Catering services advertise in many newspapers. If one newspaper flounders, this will not create problems for you.
- You ask some of your friends whether *they* know where to find the desired service. Some of your friends crashing will not prevent you from finding a caterer. The name server is distributed in this case as well, and, depending on how sociable you are, perhaps better.

Having found the address or telephone number of a catering service, you have to find a way to route your request to them. Thus, match-making between clients and services necessarily precedes routing in a mobile society. Note that the catering service, in order to execute the task you set them, may call on other services such as a car rental service. The catering service then is a client with respect to the car rental service.

Let us translate this in processing environments. The design objectives of the *Amoeba* distributed operating system project [12] motivated the design and analysis of a mathematical model for the so-called *name-server* mechanism in a distributed system with mobile processes (and objects, henceforth subsumed under processes) [13]. A *name-server* is a mechanism that translates names of processes into locations in multiprocessor networks where processes have *names* but no permanent *addresses*. This is a central part of the design of many distributed operating systems, and is analogous to the telephone system's directory assistance server: given a *name* it returns an *address*. A single *centralized* name server in the network can be taken out through a single processor crash, thereby effectively killing all communication and crashing the entire network. A more robust solution is *distributing* the name server. A great variety of options and problems of both theoretical and practical interest are attached to this issue.

More generally, we address the problem of matching mobile processes in a multiprocessor environment without centralized control. We call this *distributed match-making*. Various issues in distributed control can be thought of in terms of the distributed match-making paradigm. One of them is the name server, another one is *mutual exclusion*.

6.1.1. Name server. New generation computers must be fast, reliable, and flexible. One way to achieve this is to build them from a small number of basic processor-memory modules that can be assembled together to realize machines of various sizes. The use of multiple modules can make the machines not only fast, but also achieve a substantial amount of fault tolerance. The primary difference between machines should be the number of modules, rather than the type of the modules. In principle, any of these machines can be gracefully increased in size to improve performance by adding new modules or decreased in size to allow removal and repair of defective modules. The software running on the various machines should be in essence identical. It should be possible to connect different machines together to form even larger machines and to partition existing machines into disjoint pieces when necessary, all in a way transparent to the user level software. When a user has a heavy computation to do, an appropriate number of processor-memory modules are temporarily assigned to him. When the computation is completed, they are returned to the idle pool for use by other users. Note that in this view a *computer network* is essentially such a machine on a grand scale.

Software design for these new machines can advantageously be based on the *object model*. In this model, the system deals with abstract *objects*, each of which has some set of abstract *operations* that can be performed on it. At the user level, the basic system primitive is performing an operation on an object, rather than such things as establishing connections, sending and receiving messages, and closing connections. For example, a typical object is the file, with operations to read and write portions of it. The object model is also known under the name of 'abstract data type' [14]. A major advantage of the object or abstract data type model is that the semantics are inherently location independent. The concept of performing an operation on an object does not require the user to be aware of where objects are located or how the communication is actually implemented. This property gives the system the possibility of moving objects around to position them close to where they are frequently used. Furthermore, the issue of how many processes are involved in carrying out an operation, and where they are located, is also hidden from the user.

It is convenient to *implement* the object model in terms of clients (users) who send messages to services [15]. A service is defined by a set of commands and responses. Each service is handled by one or more server processes that accept messages from clients, carry out the required work, and send back replies.

More precisely, *services* are offered by a number of *server* processes, distributed over the network. *Client* processes send *requests* to services; the services carry out these requests and return a *reply*. Essentially, every job in the system is executed by a dynamic network of servers executing each other's requests. So a process can be a client, a server, or both, and change its role dynamically. New services can be created by installing server processes for them. Services can be removed by destroying their server processes (or by making them stop behaving like a server, i.e., by telling them to stop receiving requests). Server processes can be migrated through the network, either by actually moving the process from one host to another, or only in effect, by destroying the server process in one host and creating another one in a different host at the same time. A specific service may be offered by one, or by more than one server process. In the latter case, we assume that all server processes that belong to one service are equivalent: a client sees the same result, regardless of which server process carries out its request. A process resides in a network *node*. Each node has an *address* and we assume that, given an address, the network is capable of routing a message to the node at that address.

6.1.1.1. The problem of match-making. Before a client can send a request to a server which provides the desired service, the client has to locate that server. The problem of efficient *routing* arises at a later stage; first the address of the destination has to be found in a *match-making* phase. We can view match-making as yet another service in the system, be it the *primus inter pares*. Thus, we need to implement a *name server* to serve a connection between client process and server process.

A *centralized* name server must reside at a so-called *well-known address* which does not change and is known to all processes. (Clearly, the name server cannot be used to locate itself.) When the host of the name server crashes, the entire network crashes. This solution also causes an overload of messages in the neighbourhood of the host.

When clients *broadcast* for services with 'where are you' messages, we have an example of a *distributed* name server. This solution is more robust than the centralized one. But in large store-and-forward networks, where messages are forwarded from node to node to their destination, broadcasting is considerably more costly than sending a message directly to its destination. Broadcast messages are sent to every host, while point-to-point messages need only pass through the hosts on the path between client and server. Conventional broadcast methods for locating services need a minimum of $\Omega(n)$ message passes to do the broadcast (e.g., via a spanning tree [16,17]).

We have investigated realizations of name servers in the entire range between centralized and distributed forms. The efficiency of solutions is measured in terms of message passes and local storage. It appears that, in many n-node networks, very efficient distributed match-making between processes can be done in $O(\sqrt{n})$ message passes, by using limited numbers of point-to-point messages.

6.1.1.2. Locate algorithms. In all cases, the method to locate a service is the following. Let U be the set of nodes (i.e., processors) of the network. The network is a communication graph with two-way noninterfering communication channels between directly connected nodes. It is assumed that the nodes communicate only by messages and do not share memory. An error-free underlying communications network supports the message transfers in which the delivery time may vary but messages between two nodes are delivered in the order sent. Each of these processes is considered both a potential server (i.e. it can offer a service identified by π) as well as a potential customer (i.e. it may request a service). Let a process p reside at a host node $h(p)$. Since processes may migrate, die or be created, $h(p)$ can change, become empty or nonempty. Here we make the simplifying assumption that for the segment of time of the actual match-making instance, the process/processor allocation does not change. Location of services by the processes is achieved by the following procedure. Each server s selects a set $P(s)$ of nodes and posts at these nodes the availability of the service it offers and the address $h(s)$ where it resides. (Each node in $P(s)$ stores this information in its individual *cache*.) When a client c wants to request a service it selects a set $Q(c)$ of nodes and queries each node in $Q(c)$ for the required service. When $P(s) \cap Q(c)$ is not empty the node (or any node) in $P(s) \cap Q(c)$ will be able to return a message to c stating the address $h(s)$ at which the service is available (recall that this information is already stored in the caches of all the nodes in $P(s)$). For example, a *centralized* name-server corresponds to

$$P(s) = \{x\}, \ Q(c) = \{x\},$$

broadcasting corresponds to

$$P(s) = \{h(s)\}, \ Q(c) = U,$$

while what we may call *sweeping* corresponds to

$$Q(c) = \{h(c)\}, \ P(s) = U,$$

for all servers s and clients c with $h(s), h(c) \in U$ and some $x \in U$. Another example is the *Manhattan network*. The set U of nodes consists of pairs (i,j), with $i=1,\ldots,m$, $j=1,\ldots,n$. For all $(i,j) \in U$, a server s residing at (i,j) posts at the set

$$P(s) = \{(i,1),\ldots,(i,n)\},$$

and a client c residing at (i,j) queries the set

$$Q(c) = \{(1,j),\ldots,(m,j)\}.$$

We restrict ourselves to methods where the sets $P(s)$ and $Q(c)$ depend on the respective hosts $h(s)$ and $h(c)$ only. It therefore makes more sense to talk about $P(h(s))$ and $Q(h(c))$ instead of $P(s)$ and $Q(c)$. Thus, we define the collection of posting and querying tactics of the set of nodes U, to implement the name-server, as a single *strategy*

$$P, Q : U \to 2^U,$$

(where 2^U is the set of all subsets of U) for *match-making* in the given network.

For each set S, let $|S|$ denote the number of elements of S. The *complexity* of a match-making strategy is the *average* number of messages involved in making a match between a pair of nodes i and j, that is

$$\frac{1}{n^2} \sum_{i=1}^{n} \sum_{j=1}^{n} |P(i)| + |Q(j)|$$

In the $m \times n$ Manhattan network, above, this cost is $m+n$, i.e., $2\sqrt{|U|}$ if $m=n$.

6.1.2. Mutual exclusion. Another application of the match-making paradigm is distributed *mutual exclusion*. Consider n processes or processors which can compete for a single resource in the system, while this resource can be granted to only one of them at a time. An example is a printer which can be used by several machines or processes. The problem consists in designing a protocol which ensures that only one process is granted access to the resource at a time, while satisfying certain 'niceness' conditions such as absence of deadlock. This problem was originally formulated by Dijkstra, who also gave a first solution [18]. The assumption of the availability of mutual exclusion underlies much of the work in concurrency. For a thorough treatment see [19]. Let the network be as before. In such a distributed system, each network node can issue a mutual exclusion request at an arbitrary time, see e.g. [20]. In order to arbitrate the requests, any pair of two requests must be known to one of the arbitrators. Since these arbitrators must reside in network nodes, any pair of two requests originating from different nodes must reach a common node. Assume that each node i must obtain a permission from each member of a subset $S(i)$ of U before it can proceed to enter its critical section. Then for each pair $(i,j) \in U^2$ we must have $S(i) \cap S(j) \neq \emptyset$ so that the node in the intersection can serve as arbitrator. The complexity of a distributed mutual exclusion strategy is the average number of messages involved in a mutual exclusion request from a node i, with the average taken over all nodes. In [20] the situation is analysed where each node in the network serves as arbitrator equally often, that is, $|U|$ times. The actual algorithm presented uses at most $5 \cdot |S(i)|$ messages, where for some K, $|S(i)| = K$ for all i, $i \in U$. It is clear that at least $2K$ messages are required: K messages to query a set $S(i)$, and K answers from every member of $S(i)$ to i. The overhead of $3K$ messages arises from the necessary locking and unlocking protocols to guarantee that no more than one node can simultaneously be in the critical section, to resolve conflicts, and to prevent *deadlock* (i.e., circular waiting among the nodes requesting mutual exclusion) and *starvation* (a node which wants to enter its critical section can be prevented from doing so forever). Here, we may view a *strategy* for distributed mutual exclusion as a mapping

$$S: U \to 2^U$$

and view it as a restricted case of match-making for which the symmetry condition $P(i) = Q(i)$ ($= S(i)$) holds for all $i \in U$.

One way to achieve this symmetry is to let the functions P, Q be as before, and set $S(i) = P(i) \cup Q(i)$ for all i, $i \in U$. As an example, in the Manhattan network we obtain:

$$S(i,j) = \{(i,1), \ldots, (i,n), (1,j), \ldots, (m,j)\}.$$

The complexity of this mutual exclusion strategy turns out to be again related to $m + n$, since this is the size of $S(i,j)$, and $|S(i,j)| = 2\sqrt{|U|}$ for $m = n$.

More frugal is the example of the *projective plane*. The projective plane $PG(2,k)$ has $n = k^2 + k + 1$ points and equally many lines. Each line consists of $k + 1$ points and $k + 1$ lines pass through each point. Each pair of lines has exactly one point in common (and each pair of points has exactly one line in common). The set of $k + 1$ points incident on any of the $k + 1$ lines incident on a node $i \in U$ serves as choice for $S(i)$. Here $|S(i)| \approx \sqrt{|U|}$. This case is extensively analysed in [20].

More common in the literature are solutions with a single centralized arbiter x, that is, there is an $x \in U$ such that $S(i) = \{x\}$ for all $i \in U$. This solution is cheap in messages but vulnerable. If the host of the arbiter crashes, mutual exclusion dies as well. As a more robust and distributed solution, broadcasting is generally used. That is, $S(i) = U$ for all $i \in U$. This solution is robust but expensive. This solution is truly distributed, but far more expensive than the $\sqrt{|U|}$ solution above.

6.1.3. Results on distributed match-making. In [13,21] we developed a class of distributed algorithms for match-making between processes in computer networks. We frame this in the formalism developed above. Application of the results to distributed mutual exclusion are done by imposing the symmetry condition $P(i) = Q(i)$. E.g., assume we have a complete communication graph on n nodes. Centralized match-making corresponds in the case of the name-server to a centralized directory, and in the case of mutual exclusion to a centralized arbiter. This costs a constant number of messages per match-making instance, independent of the size of the network. Truly distributed match-making with examples in both problems we saw above. The cost here is order $\sqrt{n}$. As a last case we mention hierarchical match-making, which carries a cost of order $\log n$ messages. It can be shown, that the algorithms considered are optimal in number of message passes among all algorithms with the same measure of 'distributedness'. In general, a lower bound on the required number of messages for match-making is provided for the entire range from centralized to truly distributed methods. Lower bounds on the message complexity of distributed match-making in the multidimensional case (as opposed to the two-dimensional case considered above), and in weighted versions of the problem, are derived in [21].

6.1.4. Related work. Essentially the Manhattan topology method for implementing the name server has been used before in the torus-shaped Stony Brook Microcomputer Network [22] and a related method for distributed mutual exclusion was given in [20]. In [23] match-making is done by broadcasting. Some current multiprocessor systems avoid the communication overload due to mobile processes, which use broadcasting to do the match-making, by opting for the processes to run on fixed processors [24]. Other system designers have chosen for mobile processes, but use the crash-vulnerable solution of a centralized name server [25]. Methods which maintain a tree of forwarding addresses in the network, for each mobile process, have been used in [26]. Apart from name server and mutual exclusion, also voting schemes for version management of replicated data [27] seem related to distributed matchmaking. We have not investigated this connection.

6.2. Atomic shared register

In the distributed algorithms above we assumed that memory references to nodes where the match is made are nonconcurrent, i.e., all accesses to the node are equivalent to those accesses made in some sequence. Thus, if two or more processes simultaneously attempt to access, then the accesses will be made in some arbitrary sequential order: the effect of concurrent accesses $a1$ and $a2$ is equivalent to the sequence $a1;a2$ or the sequence $a2;a1$. This is a common assumption in treatments of concurrent processes. In the case of two simultaneous writes to a shared register, one of the values appears in the register upon completion. In the case of an overlapping read and write, the read returns the value before the write or the value after the write. Nonconcurrency of the register can be accomplished either by requiring the programming language to provide the necessary exclusive access, or by implementing the exclusion with a 'readers-writers' protocol [28]. Such an approach requires locking the register such that a write operation is never executed concurrently with any other operation, that is, a reader wait while a writer is accessing the register and vice versa. Thus, to implement mutual exclusion at the higher level we then need it at a lower level. It therefore appears that *any* method of achieving such exclusive access, whether implemented by the programmer or the compiler, requires in the end a lower-level *shared* register where reads and writes can overlap.

To retain consistency, an implementation of such a shared register has to provide the property that to an external observer concurrent memory accesses *appear* nonconcurrent and nonconcurrent accesses *appear* to preserve their order. That is, each access *appears* to take place in an indivisible grain of time, is *atomic*.

Suppose we have been able to construct an atomic flip-flop which can be tested (read) by one component (the *reader*) and set (written) by another component (the *writer*). How do we make an atomic register of n bits which can be read by one component and written by another? The writer can never write all flip-flops simultaneously, and neither can the reader read all flip-flops simultaneously. It is already a problem how to ensure that the reader gets either the new or the old value. Worse, if the register contains only values in some code, say n-bit words with k bits equal 1, how do we ensure that a read which overlaps a write does not return a word with $\neq k$ bits equal 1.

This problem is rooted in hardware design issues of concurrent accesses to registers by asynchronous components [29], and asynchronous interprocess communication [30]. Let us assume that different processors communicate using a shared memory unit. (This is the case even when the processors communicate by message passing [30].) We shall call the memory unit involved a *register*. Since physical reads and writes take time, and the processors concerned are unaware of each other's reading and writing - unless they consult another shared register - reads and writes may overlap. The problem is made worse by the fact that the communicating processors may use different technologies and operate at vastly different speeds. For instance, the writes originate from an IBM PC, and the reads originate from a Cray supercomputer. Thus, reads/writes of one processor may take many orders of magnitude more time than reads/writes of the other processor. Solving the problem of concurrent shared register access by some (possibly hardware) solution of the nature of mutual exclusion, synchronization, or execution rounds, slows everyone down to the operating speed of the slowest processor. We do not want to make any assumptions about the relative processor speeds. Therefore, we want a solution which allows all processors involved to proceed at their own pace, regardless of what the others do, and yet obtain a behaviour which *appears* serial in time.

The nature of concurrency and interaction of higher level system operation executions, consisting of many interleaved low level constituent operation executions (like the setting or testing of single flip-flops), has been beautifully analysed by Lamport [30,19]. In [30] he has constructed an atomic register, which can be read by one processor and written by one other processor. This solution starts from scratch, that is, from the most primitive asynchronous (hardware) components such as flip-flops in different technologies. We can think of such a register as a black box with two wires sticking out, a write wire and a read wire. A processor can write through the write wire, and another processor can read through the read wire, each at its own speed and performing the constituent actions of its protocol at arbitrary intervals. None of them waits for the other in any sense. Yet, as if by magic, there is way to shrink the time intervals the actions take to time *instants* (each instant inside its corresponding interval) such that the resulting *imaginary but possible* sequence of actions is consistent: a read returns the value the last previous write in the sequence wrote. That is, we imagine that the actions take place in their entirety at the precise instants the intervals have been shrunk into, and this gives a consistent result. Lamport's work raises the question of implementing an atomic multireader, multiwriter register: an atomic shared register which can be read and written by all n processors in a network. A solution to this problem, using n^2 Lamport registers in a matrix, is given in [31]. Details go far beyond the scope (and length) of this paper.

CONCLUSION

By concentrating on a few examples I have illustrated the thesis that the spirit of computing is formed by the interaction between machine architectures and algorithms. I have shown that the new non-sequential architectures of today will force the algorithm designer to consider completely new resources, such as cost of communication and wires in various forms, as determining factors for the performance or even feasibility of his algorithms for these architectures. The emerging new architectures themselves, by nature of having more degrees of freedom (not only in time, but in both space and time), call forth problems which did not arise, nor were expressible, in terms of the previous serial generation. These new problems require a new class of algorithms, which I have called 'distributed control'. Over the horizon there are ever newer architectures, with unfamiliar cost measures determining algorithm performance, and unheard of problems requiring yet other breeds of algorithms, shaping and molding computing anew.

ACKNOWLEDGEMENT

I thank Lambert Meertens for advice and help in preparing the final manuscript.

REFERENCES
1. H.H. GOLDSTINE (1972). *The Computer: from Pascal to von Neumann,* Princeton University Press, Princeton, N.J.
2. A.M. TURING (1936). On computable numbers with an application to the Entscheidungsproblem. *Proc. London Math. Soc. 42,* 230-265; Correction, *Ibid. 43* (1937) 544-546.
3. A.W. BURKS, H.H. GOLDSTINE, J. VON NEUMANN (June, 1946). *Preliminary Discussion of the Logical Design of an Electronic Computing Instrument,* Report, Princeton Institute for Advanced Study (Second Edition, September 1947).
4. J. VON NEUMANN (May, 1945). *First Draft of a Report on the EDVAC,* Draft Report, Moore School of Electrical Engineering, University of Pennsylvania, Philadelphia.
5. A. VAN WIJNGAARDEN, A.B. FRIELINK, H. VAN DE WEG, E.W. DIJKSTRA (1964). *NRMG 1959-1964; Voordrachten Gehouden op 3 April 1964,* Nederlands Rekenmachine Genootschap, Amsterdam.
6. W.D. HILLIS (1985). *The Connection Machine,* MIT Press, Cambridge, Mass.
7. C. MEAD, L. CONWAY (1980). *Introduction to VLSI Systems,* Addison-Wesley, Reading, Mass.
8. J.D. ULLMAN (1984). *Computational Aspects of VLSI,* Computer Science Press, Rockville, Maryland.
9. C. MEAD, M. REM (1982). Minimum propagation delays in VLSI. *IEEE J. on Solid State Circuits SC-17,* 773-775; Correction: *Ibid. SC-19* (1984) 162.
10. P.M.B. VITÁNYI (1985). Area penalty for sublinear signal propagation delay on chip. *Proceedings 26th Annual IEEE Symposium on Foundations of Computer Science,* 197-207.
11. S.K. TEWKSBURY, L.A. HORNAK, A. LIGTENBERG (June, 1986). *The Impact of Component Interconnects on Future Large Scale Systems,* Memorandum, AT&T Bell Laboratories, Holmdel, N.J.
12. A.S. TANENBAUM, S.J. MULLENDER (1986). The design of a capability-based distributed operating system. *The Computer Journal 29,* to appear.
13. S.J. MULLENDER, P.M.B. VITÁNYI (1985). Distributed match-making for processes in computer networks. *Proceedings 4th Annual ACM Symposium on Principles of Distributed Computing,* 261-271.
14. B. LISKOV, S. ZILLES (1974). Programming with abstract data types.

SIGPLAN Notices 9, 50-59.
15. A.S. TANENBAUM, S.J. MULLENDER (1981). An overview of the Amoeba distributed operating system. *Operating System Review 15,* 51-64.
16. Y.K. DALAL (April 1977). *Broadcast Protocols in Packet-Switched Computer Networks,* Ph.D. Thesis, Stanford University, DSL Tech. Report 128.
17. A. SEGALL, B. AWERBUCH (1983). A reliable broadcast protocol. *IEEE Transactions on Communications COM-31,* 896-901.
18. E.W. DIJKSTRA (1965). Solution to a problem in concurrent programming control. *Communications of the ACM 8,* 567.
19. L. LAMPORT (1986). The mutual exclusion problem, Parts I and II. *J. Assoc. Comp. Mach. 33,* 313-326, 327-348.
20. M. MAEKAWA (1985). A $\sqrt{n}$ algorithm for mutual exclusion in decentralized systems. *ACM Transactions on Computer Systems 3,* 145-159.
21. E. KRANAKIS, P.M.B. VITÁNYI (March, 1986). *Distributed Control in Computer Networks and Cross-Sections of Colored Multidimensional Bodies,* Technical Report MIT/LCS/TM-304, Massachusetts Institute of Technology, Laboratory for Computer Science, Cambridge, Mass.
22. D. GELERNTER, A.J. BERNSTEIN (1982). Distributed communication via global buffer. *Proceedings 1st ACM Symposium on Principles of Distributed Computing,* 10-18.
23. D.J. FARBER, K.C. LARSON (April 1972). The system architecture of the distributed system - the communication system. *Polytechnic Institute of Brooklyn Symp. on Computer Networks.*
24. CH. L. SEITZ (1985). The cosmic cube. *Communications of the ACM 28,* 22-33.
25. R.M. NEEDHAM, A.J. HERBERT (1982). *The Cambridge Distributed Computer System,* Addison-Wesley.
26. M.L. POWELL, B.P. MILLER (1983). Process migration in DEMOS/MP. *Proceedings 9th ACM Symposium on Operating Systems Principles,* 110-119.
27. D.K. GIFFORD (1979). Weighted voting for replicated data. *Proceedings 7th ACM Symposium on Operating Systems Principles.*
28. P.J. COURTOIS, F. HEYMANS, D.L. PARNAS (1971). Concurrent control with 'readers' and 'writers'. *Communications of the ACM 14,* 190-199.
29. J. MISRA (1986). Axioms for memory access in asynchronous hardware systems. *ACM Transactions on Programming Languages and Systems 8,* 142-153.
30. L. LAMPORT (1986). On interprocess communication, Parts I and II. *Distributed Computing 1* (to appear).
31. P.M.B. VITÁNYI, B. AWERBUCH (1986). Atomic shared register access by asynchronous hardware. *Proceedings 27th Annual IEEE Symposium on Foundations of Computer Science.*

RAYMOND H. FOGLER LIBRARY

DATE DUE

BOOKS ARE SUBJECT TO
RECALL AFTER